Advances in Computer Vision and Pattern Recognition

The field of computer vision and pattern recognition has a rich history of nearly 50 years. In the past decade, however, the field has experienced remarkable advances in scene understanding and image generation. This advancement is driven by three key factors: (a) the availability of large, diverse datasets, (b) the accessibility of cloud and personal computing, and (c) the open release of advanced neural network architectures and models. These breakthroughs have led to significant successes across numerous application domains, including intelligent transportation, augmented reality, healthcare, agriculture, oceanography, and more. The ACVPR book series aims to introduce, analyze, and synthesize recent and foundational research, offering valuable references to both beginners and expert practitioners. The series covers timely topics such as:

- Deep Learning for Vision
- Large-scale Foundational Models
- Generative Methods
- Multimodal Learning (vision, audio, language, action, etc.)
- Neural Fields for Vision
- 3D Computer Vision
- Computational Photography, Display and Illumination
- Video Understanding and Synthesis
- Virtual, Mixed and Augmented Reality
- Biological and Human Vision
- Physics-based Vision
- Vision for Graphics
- Ethics in Computer Vision
- Applications (Robotics, Agriculture, Health, Intelligent Transportation, Oceanography, Safety/Security, etc.)

This series includes monographs, introductory and advanced textbooks, and state-of-the-art collections. Furthermore, it supports Open Access publication mode.

Mengmeng Wang • Xiangjie Kong • Guojiang Shen

Visual Object Tracking across Modalities

Foundations, Methods, and Future Directions

Springer

Mengmeng Wang
Computer Science and Technology
Zhejiang University of Technology
Hangzhou, China

Xiangjie Kong
Computer Science and Technology
Zhejiang University of Technology
Hangzhou, China

Guojiang Shen
Computer Science and Technology
Zhejiang University of Technology
Hangzhou, China

ISSN 2191-6586 ISSN 2191-6594 (electronic)
Advances in Computer Vision and Pattern Recognition
ISBN 978-981-95-3663-4 ISBN 978-981-95-3664-1 (eBook)
https://doi.org/10.1007/978-981-95-3664-1

This Springer imprint is published by the registered company Springer Nature Singapore Pte Ltd.
The registered company address is: 152 Beach Road, #21-01/04 Gateway East, Singapore 189721, Singapore

Dedicated to our families for their support during the preparation of this book and to all our students for their ideas, motivations, and patience.

Preface

Visual Object Tracking (VOT) is a fundamental task in computer vision, essential for applications such as intelligent surveillance, autonomous driving, and robotic navigation. Over the past decades, VOT has rapidly transitioned from traditional methods based on handcrafted features to deep learning–based approaches, significantly enhancing its accuracy and robustness. Benefiting from the rapid development of deep learning and multi-modal sensing technologies, VOT is evolving to handle increasingly complex and dynamic environments. In applications like autonomous driving, where robust perception under varying weather and lighting is crucial, or in intelligent surveillance requiring reliable tracking across sensor types, multi-modal VOT plays a key role in ensuring accurate and resilient object tracking.

This book, *Visual Object Tracking across Modalities: Foundations, Methods, and Future Directions*, is designed to address these challenges. It aims to provide a comprehensive introduction to deep learning–based VOT across diverse data modalities, with a focus on how to effectively utilize both single-modal and multi-modal information to enhance tracking performance in complex environments.

Each chapter of this book is structured to stand alone, allowing readers to focus on specific areas of interest according to their needs. Chapter 1 establishes the foundational framework of VOT by introducing its definition, significance, and broad application scenarios. It outlines the core challenges that impact tracking reliability in dynamic environments. The chapter further organizes mainstream tracking methods by data modality and highlights the role of sensing diversity. It concludes by emphasizing the growing importance of multi-modal fusion for building adaptable and resilient tracking systems. Chapter 2 establishes the foundational landscape for VOT by outlining key theoretical bases and representative tracking paradigms. It aims to equip readers with essential conceptual tools for understanding visual representation, temporal modeling, and tracking architectures. The chapter emphasizes the evolution of tracking frameworks toward data-driven, unified, and end-to-end solutions. It also sets the stage for deeper technical discussions in subsequent chapters by clarifying current research directions and structural trends in the field. Chapter 3 outlines the foundational landscape of single-modal object tracking, organized by sensing modality and technical approach. It sets up a structural framework that connects representative tracking paradigms with different sensor types, highlighting methodological evolution and comparative performance. The chapter emphasizes how modality-specific characteristics shape algorithm design and influence practical challenges. The chapter's objective is to provide a systematic understanding of tracking methods and inspire further exploration within each modality domain. Chapter 4 provides a comprehensive overview of multi-modal object tracking, emphasizing the integration of complementary data sources to enhance robustness in dynamic and challenging environments. It outlines key fusion strategies and categorizes tracking methods by modality combinations and integration stages. The chapter also highlights the practical value of multi-modal approaches and points to future directions for improving efficiency, adaptability, and generalization. By establishing a unified framework, the chapter lays the groundwork for advancing the development of intelligent visual tracking systems. Chapter 5 outlines the foundational infrastructure for evaluating VOT, focusing on the dual pillars of benchmark datasets and evaluation protocols. It establishes a systematic taxonomy of datasets across sensing modalities and tracking challenges, offering a unified view of diverse tracking scenarios. In parallel, it summarizes widely adopted evaluation standards that ensure fairness, comparability, and robustness in performance assessment. This chapter thus provides the methodological backbone for empirical VOT research and future algorithm development. Chapter 6 outlines the strategic directions shaping the future of VOT, focusing on how the field is evolving to meet real-world challenges. It is structured around five key themes that aim to improve generalization, adaptability, and efficiency in tracking systems. The chapter emphasizes a shift toward more intelligent, flexible, and resource-aware tracking approaches. Its overarching goal is to guide future research and development toward scalable, practical, and robust VOT solutions.

This book is intended for researchers, practitioners, and students working in the field of VOT across multiple data modalities. It aims to serve as a comprehensive reference that bridges theoretical foundations with practical applications, offering both depth of knowledge and real-world relevance.

We sincerely hope that the cutting-edge insights and systematic knowledge in this book will provide readers with continuous inspiration and solid theoretical support in their exploration of VOT and jointly promote innovative development in this field.

Finally, we extend our heartfelt gratitude to the colleagues and students who shared their insights and feedback during the writing process. We are especially thankful to our family for their unwavering support and understanding throughout this journey.

Hangzhou, China
June, 2025

Mengmeng Wang
Xiangjie Kong
Guojiang Shen

Declarations

Competing Interests The authors have no competing interests to declare that are relevant to the content of this manuscript.

Contents

Chapter 1
Introduction

Abstract This chapter introduces visual object tracking (VOT), a core task in computer vision that involves continuously locating a target across video frames. It begins by defining the tracking objective and highlighting its importance in real-world applications such as surveillance, autonomous driving, robotics, and healthcare. The chapter then outlines key challenges faced in VOT, including deformation, occlusion, scale variation, motion blur, illumination changes, and background clutter. Following this, it categorizes tracking methods based on single and multi-modal data inputs, including RGB, thermal, LiDAR, event-based sensors, and combinations such as RGB-LiDAR or RGB language. Each modality is analyzed in terms of its strengths, weaknesses, and application contexts. Finally, the chapter emphasizes the growing importance of multi-modal fusion in enhancing robustness and adaptability under complex environmental conditions. Structured around concept definition, challenge identification, and modality-based method analysis, this chapter provides foundational knowledge for designing resilient tracking systems.

1.1 Visual Object Tracking Basics

1.1.1 Task Definition

Visual Object Tracking (VOT) aims to locate a specific object continuously in both space and time throughout a video sequence. It starts with a given bounding box in the first frame. The tracker then estimates a sequence of object states, denoted as $\{S_t\}_{t=2}^{T}$. Each state $S_t = (x_t, y_t, w_t, h_t, \theta_t)$ includes the object's center coordinates (x_t, y_t), width w_t, height h_t, and orientation angle θ_t relative to the center axis. A major challenge in VOT is combining the target's appearance features with motion patterns. This fusion must enable accurate and continuous localization under complex conditions. These conditions include deformation, occlusion, and changes in lighting. The system must also ensure real-time performance when handling these dynamic disturbances.

M. Wang et al., *Visual Object Tracking across Modalities*, Advances in Computer Vision and Pattern Recognition, https://doi.org/10.1007/978-981-95-3664-1_1

In VOT, two key concepts are frequently used. First is the template, which serves as a reference for the target object during tracking. This template can be initialized from the first frame or updated from the previous frame. In some cases, it is maintained using a memory bank that adapts over time. This strategy enables the tracker to handle changes in the target's appearance, including deformation and lighting variations. The second concept is the search region (also called the candidate region). This region defines where the tracker expects to find the target in the current frame. It is usually centered at the location predicted in the previous frame and is expanded by a margin based on the target's size.

A typical deep learning–based VOT system extracts features from both the target template and the search region. This is done using a pre-trained backbone network, such as ResNet-50 or MobileNetV3. In the tracking stage, the system predicts potential target locations through temporal modeling. Common methods include Long Short-Term Memory (LSTM) networks and Transformer-based architectures. The system then calculates similarity between the template and the candidate regions using a matching network. The resulting matched features are passed into a target localization network. This network usually contains two parts. The first is a classification head that generates a confidence score heatmap to indicate where the target is most likely to appear. The second is a confidence prediction network that refines the exact position of the target bounding box.

1.1.2 Applications

VOT technologies have become vital components in many intelligent systems. They enable reliable and continuous perception in dynamic environments. With the advancement of deep learning and multi-sensor fusion, modern VOT methods can now address complex visual challenges. These include occlusion, object deformation, and large-scale appearance changes. VOT systems are widely applied in areas such as public safety, transportation, healthcare, industrial automation, robotics, and human–computer interaction. Their versatility is driving both innovation and real-world deployment. The following six representative paradigms demonstrate how VOT techniques contribute to practical applications across these domains.

1. Video Surveillance and Security Systems
VOT plays a crucial role in modern surveillance and security systems. It enables continuous monitoring of individuals or objects in dynamic environments. VOT supports cross-camera tracking, allowing targets to be followed seamlessly as they move between different camera views. It also aids in anomaly detection by recognizing suspicious behaviors. These include loitering, unattended baggage, or abnormal crowd gatherings. Such capabilities help generate timely alerts for security personnel. In large public events or transportation hubs, VOT can work with crowd density estimation algorithms to produce real-time heatmaps. These visualizations support effective resource allocation and risk prevention. With the progress of

deep learning and edge computing, VOT systems have become more robust and generalizable. As a result, they are now key components of intelligent security infrastructure.

2. Autonomous Driving and Environmental Perception
In autonomous driving systems, VOT is a core component for environmental perception and path planning. It enables real-time tracking of dynamic objects such as vehicles, pedestrians, and bicycles. This provides continuous and stable state information for the perception module. When integrated with multi-sensor fusion strategies—including Light Detection and Ranging (LiDAR), radar, and camera data—VOT improves system reliability across different lighting conditions, weather situations, and complex traffic environments. In urban areas, for instance, VOT helps track pedestrians who suddenly cross the road. This supports emergency braking systems. On highways, it assists in predicting lane-changing behavior of nearby vehicles, which benefits adaptive cruise control. VOT also contributes to motion trajectory modeling of surrounding objects. This helps the system understand the behavior of traffic participants, enhancing both safety and intelligence in autonomous driving.

3. Medical Imaging and Surgical Assistance
In medical imaging and minimally invasive surgery, VOT is widely used to monitor surgical instruments, organ tissues, and anatomical landmarks in real time. With high-precision tracking, surgeons receive intuitive guidance during procedures, which improves both accuracy and safety. For example, in endoscopic surgery, VOT captures the three-dimensional motion of surgical tools and overlays their trajectories onto preoperative organ models. This helps surgeons avoid critical vessels and nerves. In radiation therapy, VOT tracks patient respiratory motion, allowing treatment devices to adjust beam direction dynamically. This reduces radiation exposure to healthy tissue and improves targeting accuracy. VOT is also integrated into robotic surgical systems. It provides stable spatial localization and supports remote or automated operations, driving medical technology toward greater levels of intelligence.

4. Industrial Automation
VOT has important applications in industrial automation and logistics systems. On manufacturing production lines, VOT is used for tracking products on conveyor belts, monitoring assembly processes, and improving workflow coordination. In cases such as automotive parts production, electronic device assembly, or food packaging, VOT continuously identifies and tracks moving components. This ensures accurate transfer between processing stages and prevents issues like misalignment, missed assembly, or redundant operations. In addition, when combined with image recognition algorithms, VOT can detect appearance defects, size deviations, or labeling errors in products. Upon detecting anomalies, the system can trigger automatic rejection or alert mechanisms. These functions greatly improve the efficiency and consistency of quality inspection.

5. Robotic Perception and Navigation
In various robotic systems, VOT is a core technology for autonomous perception and navigation. By continuously tracking targets such as people, objects, or designated areas, robots can better understand their surroundings and make informed decisions. VOT supports essential functions like target following, path planning, and obstacle avoidance. It is often integrated with Simultaneous Localization and Mapping (SLAM) and advanced path planning algorithms to improve robot adaptability in unfamiliar or unstructured environments. For example, in service robots, VOT enables user identification and following for personalized assistance. In industrial inspection robots, it helps recognize and track key equipment to improve detection efficiency. In special-purpose robots—such as those used in rescue operations—VOT, combined with other sensing modules, supports fast target recognition and continuous tracking. This enhances task continuity and reliability. As algorithms advance and hardware becomes more powerful, VOT is playing an increasingly flexible and efficient role in robotic applications.

6. Augmented Reality and Human–Computer Interaction
In augmented reality (AR) and human–computer interaction, VOT is a key technology that enables seamless integration of virtual content with the real world. By accurately tracking physical objects, hand gestures, or head poses, VOT provides stable alignment between virtual and real elements. This improves user immersion and interaction quality. For example, in education and training, VOT tracks teaching objects and overlays virtual annotations to make learning more intuitive. In remote collaboration, it follows user gestures to enable natural interactions. In industrial design and maintenance, VOT supports real-time recognition and annotation of equipment, helping technicians quickly access operation guidance. Additionally, VOT is often combined with other computer vision techniques, such as semantic segmentation and pose estimation. This combination expands the functionality of AR applications and opens new possibilities in consumer electronics, smart manufacturing, and digital twin technologies.

1.2 VOT Key Challenges

VOT is a fundamental problem in computer vision. Despite significant progress in recent years, many challenges remain in practical applications. These challenges are especially prominent in complex environments or when integrating multi-modal information. They affect both tracking accuracy and stability. Additionally, they increase the demands on real-time performance and algorithm adaptability. The following section reviews the main issues faced by current VOT systems in real-world scenarios. It also discusses how these challenges impact tracking performance based on recent research findings.

1.2.1 *Deformation and Non-Rigid Motion*

Visual tracking of deformable and non-rigid objects remains a persistent challenge in real-world applications, as shown in Fig. 1.1. Unlike rigid targets, which maintain a stable shape during motion, flexible or soft objects undergo significant shape changes. These changes include stretching, bending, folding, and sometimes partial occlusion. Such variations make it difficult for traditional tracking algorithms to maintain consistent localization.

The core difficulty is the strong coupling between global motion and local structural deformation. As the object moves, its appearance can change drastically. This happens due to nonuniform shape variations, texture distortions, and self-similarity across different regions. These factors often cause boundary ambiguity and feature instability. They are major reasons for tracking drift or failure.

Many conventional tracking methods [3, 40, 55] rely on fixed feature points or contour models. These methods assume shape consistency over time. However, this assumption fails when tracking highly deformable targets. Such targets exhibit continuous and unpredictable appearance variations. Some adaptive approaches based on deep learning [5, 44] have improved robustness to deformation. Still, they struggle to distinguish meaningful deformation from noise, especially in long-term tracking scenarios.

Designing a tracking framework that balances sensitivity to deformation and stability in motion estimation remains a key research direction. Addressing this issue

Fig. 1.1 Example of deformation and non-rigid Motion [12]

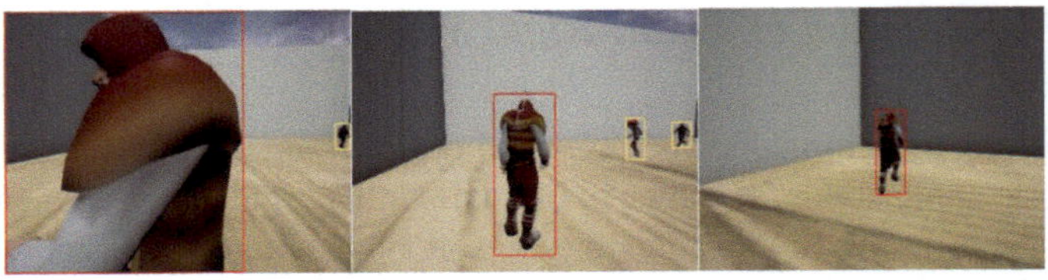

Fig. 1.2 Example of scale variation [93]

is crucial for improving performance on flexible objects, such as plastic bags. It is also important for enhancing the generalization ability of visual tracking systems across diverse applications.

1.2.2 Scale Variation

Dynamic changes in the scale of a target object pose typical challenges for visual tracking systems. As shown in Fig. 1.2, the target marked by the red box can quickly shrink or enlarge within the frame. It is also affected by similar background distractions. In drone aerial photography, rapid camera movement causes ground targets to grow quickly in size. This makes it hard for traditional fixed-size search windows to capture target boundaries accurately. In autonomous driving, small-scale pedestrians in the distance are often confused with streetlights, traffic signs, or other objects due to missing details [22].

To address this challenge, the literature [34] proposes hierarchical feature fusion to dynamically perceive target size. However, under extreme scales—such as continuous observation of cell division under a microscope—semantic association still weakens. Additionally, when the target and interfering objects have similar scales (e.g., adjacent vehicles at normal scale in the image center), tracking drift risk increases. The algorithm must balance cross-scale feature extraction with background suppression dynamically. This requirement raises higher challenges for the real-time computing architecture design of embedded devices.

1.2.3 Motion Blur

Image blur caused by fast-moving targets or cameras significantly challenges the stability of visual tracking systems. As shown in Fig. 1.3, blur can vary from mild motion artifacts to severe halo diffusion. All these effects degrade the visual information quality used for tracking [32]. In high-speed scenarios like drone tracking or autonomous emergency braking, rapid movement causes trailing effects. These effects severely distort texture details and blur edge contours. For example,

Fig. 1.3 Examples of motion blur [31]

when a drone tracks an out-of-control vehicle, a strong blur may cause the tracker to mistake the headlight trail for the vehicle's body.

In industrial applications, high-speed visual inspection systems often fail to locate or grasp components accurately due to blurred surface features [57]. Traditional template-based tracking methods are especially vulnerable to spatial correlation loss in blurred regions. Deep learning–based trackers partially address this issue by using temporal feature fusion [33]. However, the combined effects of blur and noise in low-light conditions, such as nighttime traffic monitoring, can still cause feature confusion and long-term tracking drift.

Severe blurring destroys critical visual features. This forces the system to rely more on motion prediction models to compensate. As a result, demands on real-time performance and computational resources increase. Current research focuses on jointly optimizing blur perception and motion modeling. However, balancing artifact suppression with feature fidelity during deblurring remains a key challenge in practical applications.

1.2.4 Illumination Variations and Low-Light Conditions

Target tracking in complex lighting conditions is a core challenge for visual systems in real-world scenarios. As shown in Fig. 1.4, traditional visible light sensors (first row of RGB images) often fail to track targets under strong backlighting, at night, or in dark environments. This failure occurs because texture details are easily obscured. In contrast, thermal imaging (second row of thermal images) can capture thermal radiation emitted by targets, even in the dark. However, thermal sensors have limited ability to distinguish between interfering objects with similar temperature distributions [102].

(a)

Fig. 1.4 Examples of illumination variations and low-light conditions [47]

This issue is especially prominent in nighttime autonomous driving scenarios. Headlight glare blurs the contours of vehicles ahead. Thermal imaging can detect pedestrians' heat signatures but struggles to determine their movement intentions. In security surveillance, window reflections caused by strong light and low signal-to-noise ratios in shadowed areas reduce the reliability of cross-camera target re-identification [91]. Mainstream visible-light-based trackers typically rely on color and texture features, which are prone to failure. Multi-modal fusion solutions, such as joint RGB-thermal imaging models, effectively reduce the limitations of single sensors [1]. However, aligning heterogeneous data in space and time and managing increased computational demands pose significant engineering challenges.

Sudden lighting changes, such as brightness switching at tunnel entrances and exits, cause mismatches between the tracking model and the target. This forces the system to frequently reset initialization parameters, severely disrupting tracking continuity. Current research mainly focuses on adaptive lighting-invariant feature extraction and lightweight multi-modal fusion framework design. However, achieving real-time and robust tracking on resource-constrained edge devices remains a major bottleneck for industrial applications.

1.2.5 Background Clutter

Complex background clutter is a primary cause of visual tracking failure in practical applications. As shown in Fig. 1.5, targets often get confused due to high similarity in color and texture with the surrounding environment. Examples include camouflaged military equipment in deserts or pedestrians in crowded scenes. In autonomous driving, background clutter like roadside billboards and reflective guardrails can cause trackers to mistake high-contrast areas for targets.

Fig. 1.5 Example of background clutter [70]

Fig. 1.6 Examples of occlusion [22]

In agricultural drone monitoring, periodic crop textures—such as the repeated rows in cornfields—can interfere with tracking agricultural machinery [65].

Traditional methods based on color histograms or edge features [13, 40] lack semantic understanding. They are prone to template drift in complex backgrounds, where the tracking window gradually shifts to similar background areas. Deep learning methods improve interference resistance by introducing advanced semantic features [54]. However, they still struggle when targets and backgrounds are extremely similar, such as chameleons blending with their environment. For example, in security surveillance, traditional correlation algorithms have difficulty maintaining continuous target trajectories due to local feature similarity [8]. Current research mainly focuses on decoupling background suppression from target representation [41, 48]. Still, achieving real-time operation while ensuring sensitivity and robustness to complex backgrounds remains a key challenge for VOT.

1.2.6 Occlusion and Target Appearance Changes

Dynamic occlusion and unpredictable changes in target appearance commonly cause failures in visual tracking systems. As shown in Figs. 1.6 and 1.7, targets can be partially or completely occluded in complex environments. Their appearance may also change due to lighting or posture variations. This issue is especially significant in surgical robot operations. When the endoscopic view is blocked by blood or surgical instruments, traditional fixed-template trackers often fail to locate the target. These trackers cannot adapt to sudden appearance changes, which poses potential operational risks [58].

Previous studies, such as [60], have addressed partial occlusion by dynamically updating templates. However, prolonged occlusions—like pedestrians entering an

Fig. 1.7 Examples of target appearance changes [63]

elevator and becoming invisible for a sustained time—can still cause tracking drift due to accumulated errors. The multi-hypothesis prediction tracking framework [64] generates multiple candidate trajectories during occlusion to improve robustness. Yet, in complex scenarios such as crowded environments, the number of hypotheses grows exponentially. This causes high computational overhead and limits practical use.

In practical applications, maintaining target identity continuity during occlusion is a key challenge. It requires balancing conservatism and adaptability in updating the appearance model. This balance is crucial for improving tracking generalization across scenarios. For example, in autonomous driving, when a front vehicle is fully occluded by a large truck, the system must handle sudden changes in the vehicle's position and scale once occlusion ends. It must also quickly detect potential lane-cutting vehicles in adjacent lanes. These tasks place higher demands on the tracking model's memory and prediction abilities.

1.3 VOT Across Data Modalities

1.3.1 RGB-Based VOT

RGB-based VOT uses color information from standard video sequences to locate objects. As the most common perception modality in computer vision, RGB tracking has made significant progress thanks to deep learning advancements. This section analyzes its advantages and limitations. It also discusses practical applications combined with the latest tracking frameworks. Figure 1.8 shows a typical example of RGB tracking.

Fig. 1.8 Example of rgb tracking [46]

1. Advantages

RGB single-modal tracking relies on images captured by visible light cameras. Due to its technical maturity, low hardware cost, and easy data acquisition, it is the most widely used tracking solution in industry today. Existing algorithms achieve sub-pixel localization accuracy in stable lighting and controlled backgrounds. They leverage image texture, color, and spatial context through end-to-end deep learning frameworks, such as the Transformer architecture [29]. For example, sequence learning frameworks based on temporal modeling, like SeqTrack [11], encode states from past frames to model target motion patterns. This improves robustness for fast-moving targets. Lightweight designs, such as MixFormerV2 [15], use feature extraction and attention pruning to enable real-time tracking on embedded devices. Moreover, the rich semantic content of RGB data suits scenarios requiring high visual consistency, such as human–computer interaction and video editing. For instance, virtual object insertion in film production depends on high-precision pixel-level alignment to achieve seamless visual fusion.

2. Limitations

In complex environments, RGB tracking faces inherent limitations due to multi-source interference. First, sudden lighting changes—such as entering or exiting tunnels—can distort color features. Traditional models relying on color histograms are prone to drift in these conditions [97]. Second, target deformation, occlusion, and cluttered backgrounds disrupt visual continuity. Even with memory-enhanced methods like global representation memory banks [104], long-term tracking errors still accumulate. For example, in crowded scenes, trackers based on local feature matching, such as ODTrack [101], often suffer identity switches due to similar-looking targets. When facing severe occlusions, models relying on historical state guidance, like DropMAE [81], may experience tracking interruptions from memory injection errors. Additionally, RGB data are sensitive to adverse conditions such as low light and haze. This limits its use in all-weather scenarios and calls for integration with other sensors.

3. Practical Applications

Despite some limitations, single-modal RGB tracking has been widely applied due to its low deployment cost and strong versatility. In autonomous driving, real-time monitoring of road users' motion trajectories with multi-object tracking frameworks like VideoTrack [85] supports critical functions such as collision warning. In security surveillance, lightweight trackers—such as the hierarchical Transformer designed by Kang et al. [39]—can run on edge cameras for cross-camera association of suspicious targets. In consumer electronics, foreground tracking methods (e.g., explicit visual guidance strategies [73]) improve smartphone video stabilization by separating subject motion from hand shake. RGB tracking also enables facial recognition, portrait enhancement, and gesture interaction in smartphones and tablets. These features support autofocus, real-time augmented reality rendering, and interactive experiences based on visual cues. In entertainment broadcasting, RGB tracking aids virtual background segmentation and dynamic camera repositioning, offering a cost-effective solution for live studio effects. In industrial quality inspection, object representation learning methods like those of Cai et al. [6] reliably track assembly processes and trigger defect detection when anomalies appear. In industrial automation, RGB tracking ensures precise part alignment and defect detection on manufacturing lines, maintaining stable positioning even amid mechanical vibration or partial occlusion. In agriculture, RGB-based UAV monitoring systems track crops spatially to assess growth, identify pests, and detect irrigation issues using color and shape cues. Combined with segmentation, these systems support precision farming through plant-level interventions. In medical imaging, digital pathology benefits from RGB tracking to follow tissue sample evolution across temporal slices. This helps analyze disease progression, such as observing cancer cell migration under chemical stimuli to evaluate therapies. Current research focuses on lightweight models—like separable attention mechanisms [29]—and enhancing cross-scenario adaptability through generalized relationship modeling [28]. Overall, RGB tracking is essential for high-resolution visual understanding under standard lighting conditions.

1.3.2 Thermal Infrared VOT

Thermal infrared (TIR) tracking uses the thermal radiation emitted by targets to achieve stable localization under low-light or complex lighting conditions. This section analyzes the advantages, limitations, and practical applications of single-modal TIR tracking systems. Figure 1.9 shows a typical example of TIR tracking.

1. Advantages

Thermal infrared imaging offers significant advantages in complex environments such as low light, smoke, and haze. It achieves stable tracking by capturing the target's thermal radiation. Unlike visible light, which depends on reflected environmental light, thermal imaging detects an object's infrared energy directly.

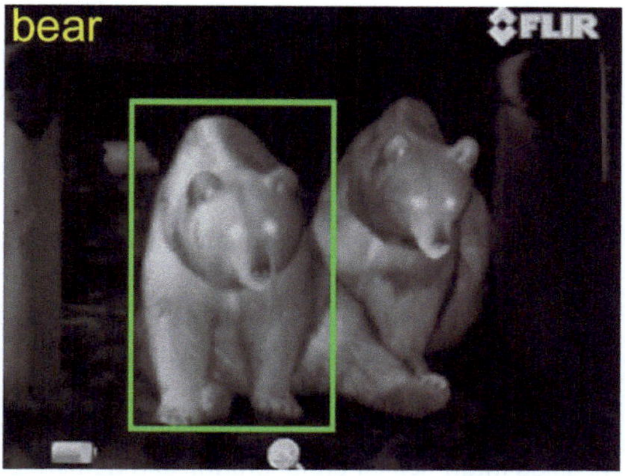

Fig. 1.9 Example of thermal infrared tracking [53]

Therefore, in scenarios like nighttime surveillance and forest fire warning, it can penetrate darkness or partial obstructions (e.g., smoke) to continuously track targets such as fire hotspots or human heat signatures [49]. Algorithmically, thermal imaging adapts better to changes in target appearance, such as camouflage or posture deformation. For example, a dual-layer feature learning framework [53] effectively handles local temperature variations caused by environmental heat exchange by combining shallow-layer temperature distributions with deep semantic features. Additionally, in military applications (e.g., aircraft exhaust plume tracking) and industrial inspection (e.g., circuit board hotspot localization), thermal infrared tracking provides unique advantages by directly reflecting the physical state of the target [68].

2. Limitations

The core challenge of thermal imaging tracking lies in its physical limitations. First, thermal infrared images generally have lower resolution and blurred texture details compared to visible light images. This makes it difficult to distinguish targets with similar shapes and temperatures, such as engine areas in densely parked vehicles [94]. Second, many non-target heat sources in the environment cause interference. Examples include ground heated by sunlight or heat from air conditioner units. In complex urban thermal scenes, traditional temperature-threshold segmentation methods often produce false positives or fail to identify targets [10]. Additionally, thermal imaging is sensitive to temperature differences within targets. For instance, thick clothing can weaken a person's thermal radiation signal, causing the tracker to lose the target [16]. Algorithmically, the generally low contrast in thermal images means traditional correlation filters based on grayscale features are easily affected by background thermal noise, leading to poor tracking stability. In contrast, deep learning methods require feature extraction modules sensitive to temperature gradients to improve thermal signal identification and characterization [92].

3. Practical Applications
Thermal imaging tracking has shown irreplaceable value in many fields. In military security, infrared warning systems with spatiotemporal feature fusion continuously track low-altitude aircraft thermal signals, overcoming visible-light detection limits [68]. In power line inspection, drones equipped with thermal cameras scan high-voltage lines to automatically identify faults such as local insulator overheating, improving automation and accuracy [82]. In medical settings, non-contact temperature monitoring systems use attribute-guided fusion networks to track temperature changes in multiple targets, applied in airport fever screening [84]. In industrial automation, thermal imaging monitors equipment in high-temperature, high-risk environments. For example, thermal trajectory analysis predicts wear trends of mechanical parts in metallurgical production lines [92]. Current trends focus on multi-sensor collaboration and lightweight deployment to promote thermal tracking in consumer devices like smart-home night monitoring [43]. Thermal infrared tracking excels in low-light, nighttime, or visually complex environments. Unlike RGB sensors, thermal cameras detect emitted infrared radiation, sensing temperature rather than reflected light. This makes them effective in surveillance, navigation, and medical diagnostics. In security monitoring, thermal tracking is widely used in border patrol, night drones, and infrastructure protection. It detects humans and animals in total darkness, through smoke or fog, and in camouflaged settings. Heat signatures remain robust despite scene clutter and visual ambiguity. In intelligent transportation, vehicle-mounted thermal cameras enhance driving safety under adverse weather or low-light conditions by tracking pedestrians, cyclists, and nearby vehicles via their thermal outlines. In wildlife monitoring, thermal tracking enables non-invasive observation of nocturnal species and temperature-sensitive behaviors. Clinically, breast cancer screening systems use thermal tracking to detect subtle changes in blood flow and tissue temperature that may signal disease.

1.3.3 LiDAR-Based VOT

LiDAR-based tracking has become dominant in outdoor autonomous systems due to its unique spatial perception capabilities. This section analyzes the advantages, limitations, and practical applications of LiDAR point cloud tracking. These insights are supported by recent advances in the field. Figure 1.10 shows an example of LiDAR-based tracking.

1. Advantages
In complex dynamic scenes, LiDAR shows strong robustness through precise 3D point cloud modeling. Unlike visual sensors that depend on lighting and texture, LiDAR is unaffected by lighting changes, shadows, or surface texture variations [99]. For example, in autonomous driving, visual systems struggle in strong backlighting or when entering and exiting tunnels. However, LiDAR relies on point cloud structure to stably perceive objects ahead. Recently, motion

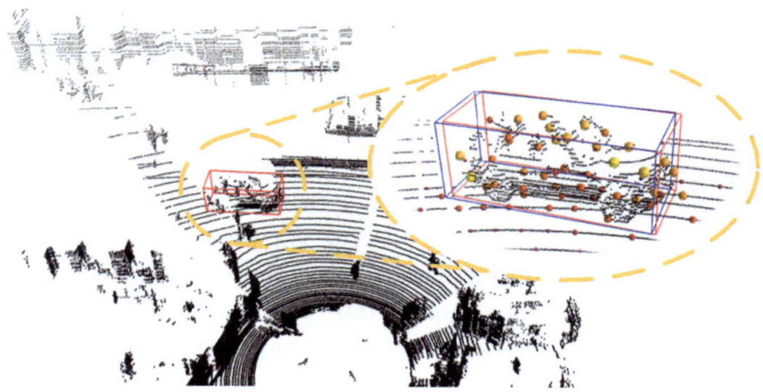

Fig. 1.10 Example of liDAR-based tracking [71]

modeling methods based on attention mechanisms, such as SiamMo [89], effectively distinguish rigid motion differences between objects and backgrounds. They achieve centimeter-level positioning accuracy even in dense traffic. Additionally, LiDAR's active ranging maintains good depth resolution for long-range tracking. In large-scale aerial monitoring, LiDAR helps drones continuously track ground-moving targets [62]. Multi-frame point cloud temporal fusion techniques like M3SOT [52] improve trajectory continuity under occlusion. This enhances adaptability in dynamic environments.

2. Limitations

The main bottleneck in LiDAR tracking is the sparsity of point cloud data and dynamic noise interference. In high-speed scenarios, such as sharp vehicle turns, the spatial sampling frequency may be insufficient. This causes blurred geometric features on the target surface, affecting accurate recognition. Traditional feature extraction methods based on voxel partitioning, like VoxelTrack [56], often suffer from bounding box offset issues [95]. Local deformations from non-rigid targets—such as swinging pedestrian arms or fluttering flags—disrupt point cloud consistency. This makes it hard for algorithms to rely on appearance features and forces them to focus more on motion prediction [98]. Algorithmically, the unordered nature and varying density of point clouds challenge the use of traditional 2D convolutional networks. Graph neural networks, such as PTTR [103], model point relationships but still lack real-time performance. In practical deployment, the high computational and storage demands of point cloud processing place strict requirements on edge device capabilities [38]. This poses a significant challenge for LiDAR tracking implementation.

3. Practical Applications

LiDAR tracking has become the standard solution in autonomous driving and intelligent robotics. In autonomous driving perception systems, motion-based tracking frameworks like P2B [67] estimate real-time motion states of vehicles and pedestrians by matching point clouds with prediction boxes. When combined with multi-sensor calibration error compensation [80], lateral position errors remain within reasonable limits. In drone inspection, lightweight one-stream networks such as EasyTrack [20] compress point cloud features, enabling embedded platforms to continuously track power line insulator defects from the air. In military applications, LiDAR integrates with inertial navigation systems like MBPTrack [87], enabling covert tracking of armored ground targets via motion pattern recognition in GPS-restricted areas. Current trends focus on cross-modal knowledge distillation, including 2D and 3D joint pre-training [80], and sparse point cloud completion techniques such as quality-aware shape reconstruction [95]. These advances drive LiDAR tracking expansion into intelligent logistics (e.g., autonomous vehicle cargo tracking) and urban digital twins (e.g., traffic flow simulation) [61]. Additionally, LiDAR tracking supports urban surveillance by monitoring crowd density and vehicle movement, aiding city planners in infrastructure optimization. In construction and logistics, it ensures spatial coordination of machinery and inventory. For structural monitoring and cultural heritage preservation, LiDAR detects deformations or shifts in buildings and archaeological sites with centimeter-level accuracy.

1.3.4 Event-Based VOT

Event-based vision has emerged as a revolutionary paradigm for high-speed tracking. It leverages microsecond temporal resolution and ultralow latency to overcome the limits of conventional frame-based sensing. This section examines the advantages, limitations, and applications of event-based tracking. The discussion is supported by recent advances in tracking frameworks. Figure 1.11 illustrates a typical event-based tracking pipeline.

1. Advantages

Event-based tracking uses the asynchronous event streams from dynamic visual sensors (DVSs). It shows unique advantages in high-speed motion and extreme lighting changes. Unlike traditional frame-based cameras, event cameras only react to pixel brightness changes. They offer microsecond-level temporal resolution and

(a) (b) (c)

Fig. 1.11 Example of event-based tracking [96]

a high dynamic range of over 140 dB. This allows precise capture of fast object motions, such as bullet trajectories or insect wing vibrations, avoiding motion blur [26]. Event cameras also consume low power and have high energy efficiency. For example, the SCTN model [36] achieves milliwatt-level power use with a spiking neural network. This makes them well-suited for embedded platforms like drones and wearable augmented reality devices. Moreover, event data are sparse, recording only changed regions. This reduces computational load and enables real-time processing of high-speed targets, such as race car trajectory prediction [96].

2. Limitations

The main bottleneck in event-driven tracking is adapting data characteristics and algorithms. Due to the sparse nature of event stream data and the lack of texture information, targets are often lost in static or slow-moving scenes. This occurs because there are too few triggering events [100]. Event data are an asynchronous spatiotemporal point cloud, structurally different from traditional frame-based images. This difference makes it difficult to apply traditional processing methods directly. Therefore, specialized spatiotemporal encoders, such as the pulse Transformer in EvTracker [96], are needed to extract motion information. However, hardware acceleration support for these models is still limited. Noise events in complex backgrounds—like random triggers from wind-blown leaves—also interfere with recognizing target motion patterns. Existing anti-interference methods, such as the suppression module by Fu et al. [25], still suffer from higher false positive rates in dynamic, dense scenes.

3. Practical Applications

Event-driven tracking uses asynchronous data streams from DVS. It offers unique advantages in high-speed motion and low-light environments, marking a paradigm shift in visual perception. In autonomous driving, event-stream-based obstacle detection systems, like HDETrack V2 [77], can stably track vehicles and pedestrians under strong sunlight or at tunnel entrances due to millisecond-level response speed. In drone search and rescue, event cameras combined with spiking neural networks (e.g., SDTrack [72]) locate vital signs by detecting thermal radiation changes at night or in smoky conditions. For industrial inspection, asynchronous tracking frameworks such as the adaptive time surface model [9] enable real-time defect localization on fast assembly lines. Integration of neuromorphic computing and event tracking, demonstrated by MambaEVT [78], supports low-latency interaction between brain–computer interfaces and biomimetic robots. Unlike frame-based cameras that capture full images at fixed intervals, event cameras record only pixel-level brightness changes. This yields microsecond latency and a wide dynamic range. In robotics, event-based tracking aids agile navigation of drones and ground robots in dynamic or cluttered environments. Traditional RGB trackers often suffer from motion blur or latency in obstacle-rich or high-speed scenarios. Event trackers provide fast, sparse updates for real-time obstacle detection, SLAM, and autonomous control with lower computational load. Aerospace applications benefit from event cameras' reliable performance in low-light and high-contrast scenes, critical for spacecraft docking and satellite maneuvers where conventional imaging fails. In competitive sports, event-based systems track joint angles and limb trajectories during fast actions like table tennis or badminton, enabling real-time biomechanical analysis and motion feedback for training. In industrial quality inspection, event tracking detects ultra-short phenomena such as sparks or micro-vibrations, identifying defects invisible to standard imaging. Event cameras also advance neuroscience and biology. For example, insect behavior analysis platforms use event sensors to decode millisecond-scale courtship dances or escape reflexes, too fast for traditional video. These insights contribute to scientific understanding and bio-inspired robotics development.

1.3.5 RGB-LiDAR Tracking in VOT

RGB-LiDAR fusion achieves millimeter-level 3D tracking by combining color texture from RGB images with spatial information from LiDAR point clouds. This synergy is vital for autonomous vehicles, enabling real-time obstacle tracking. It also supports drones in precision navigation within dynamic environments. This section reviews the advantages, limitations, and practical applications of RGB-LiDAR tracking. Recent advances in tracking frameworks are discussed. Figure 1.12 shows a typical example of RGB-LiDAR tracking.

1. Advantages

The multi-modal fusion of RGB and LiDAR greatly enhances tracking robustness in complex dynamic scenes. It leverages the complementary sensing features of both modalities. RGB cameras provide high-resolution texture details, such as vehicle color and pedestrian clothing. LiDAR offers precise 3D point cloud modeling, including distance and speed estimation. This combined multidimensional feature

Fig. 1.12 Example of rgb-lidar tracking [2]

representation is especially effective in adverse weather like rain and fog. It also performs well under sudden lighting changes, such as at tunnel entrances and exits. LiDAR's active detection avoids the visual degradation that affects RGB sensors [2]. Task-driven fusion strategies, like the resource optimization framework by Samal et al. [69], enable low-power real-time tracking on embedded systems. For example, autonomous vehicles use point cloud density-based attention to prioritize nearby high-risk targets. Meanwhile, RGB semantic data helps identify traffic signs. In multi-object intersections, such fusion methods achieve excellent trajectory prediction accuracy [35].

2. Limitations

The complexity of multi-modal data alignment and hardware constraints remain core bottlenecks limiting the widespread application of RGB-LiDAR fusion technology. Temporal and spatial asynchrony between LiDAR and RGB sensors—such as differences in scanning frequency and calibration errors of external parameters—can cause mismatches in the spatial projection of moving targets. Traditional interpolation compensation methods based on timestamps are prone to accumulating errors in high-speed dynamic scenarios [19]. Although solid-state LiDAR has improved integration levels, its limited vertical field of view and sparse point clouds (e.g., insufficient sampling of distant objects) continue to challenge precise target recognition [66]. Moreover, the computational load from heterogeneous data processing—including point cloud voxelization and image convolution—in multi-modal fusion models demands significant onboard computing resources. Embedded system optimization techniques proposed by Sualeh et al. [74] can reduce resource consumption but often trade off tracking accuracy.

3. Practical Applications

This technology is applied in three primary scenarios: autonomous driving perception systems, drone formation collaborative operations, and industrial logistics. In autonomous driving perception systems, a multi-frame feature alignment tracking framework [19] achieves stable tracking of vehicles and pedestrians within a 100-meter range by performing motion compensation on LiDAR point clouds combined with RGB temporal features, maintaining lateral positioning errors within 0.2 meters. For drone formation collaborative operations, RGB-LiDAR fusion systems—such as the multi-object tracking architecture proposed by Wang et al. [76]—utilize three-dimensional geometric features to effectively eliminate interference from visually similar targets, like vehicles reflected on glass curtain walls in high-reflection building complexes. In industrial logistics scenarios, AGV forklifts integrate solid-state LiDAR with visual semantic segmentation to realize millimeter-level localization and collision avoidance of pallet stacking states in dynamic warehouse environments, greatly enhancing loading and unloading efficiency and safety [74]. Current research trends focus on integrating neuromorphic sensors and pulse computing architectures to overcome the energy efficiency limitations inherent in traditional von Neumann architectures.

Fig. 1.13 Example of rgb-depth tracking [42]

1.3.6 RGB-Depth Tracking in VOT

RGB-Depth multi-modal fusion significantly enhances target tracking robustness in complex scenarios by combining color–texture information from RGB images with three-dimensional spatial perception from depth sensors. This section analyzes the advantages, limitations, and practical applications of RGB-Depth tracking, supported by recent advances in tracking frameworks. Figure 1.13 presents a typical example of RGB-Depth tracking.

1. Advantages
RGB-Depth tracking significantly improves target distinguishability and tracking robustness in complex scenes by integrating texture information from visible light images with the geometric perception capabilities of depth sensors. Depth information effectively mitigates tracking drift issues common in traditional RGB tracking caused by color similarity (e.g., identical-colored products on densely packed shelves) or challenging lighting conditions (e.g., low-light nighttime environments). For instance, algorithms based on adaptive depth-invariant models [83] can robustly lock onto dynamic occlusions—such as specific individuals within a crowd—by leveraging the geometric contours of targets. The three-dimensional spatial information provided by depth data further aids in accurate scale estimation and pose inference, enabling millimeter-level positioning precision in robotic grasping tasks [86]. Moreover, the depth channel's capacity to suppress background noise (e.g., differentiating targets from walls with similar textures) significantly reduces false detection rates. In particular, for multi-target tracking, optimizing trajectory association via spatiotemporal consistency constraints demonstrates clear advantages [59].

2. Limitations
The performance of RGB-Depth tracking is constrained by the inherent physical limitations of depth sensors and the challenges of effective multi-modal

fusion. Mainstream depth sensing technologies—such as time-of-flight (ToF) and structured light sensors—are susceptible to depth data loss or noise interference when exposed to strong ambient light, long-range distances, or transparent and reflective surfaces like glass walls [88], which reduces tracking reliability in outdoor environments. Moreover, temporal and spatial asynchrony between RGB and depth streams complicates multi-modal feature alignment, potentially leading to fusion inaccuracies. Although Transformer-based cross-modal interaction frameworks [27] have alleviated some alignment issues, their substantial computational cost hinders real-time application. Additionally, the relatively high power consumption and hardware expense of depth sensors limit their deployment in consumer electronics such as smartphones and drones [7].

3. Practical Application

RGB-Depth tracking technology has matured into practical solutions across robot navigation, augmented reality, and smart security. Service robots use real-time depth perception devices (e.g., Kinect V2) to avoid obstacles in dynamic home environments and continuously track users [105]. Augmented reality headsets combine SLAM with RGB-Depth tracking to anchor virtual objects persistently in physical space. This enables effects like virtual–real occlusion handling in furniture preview applications. In industrial quality inspection, depth-assisted defect localization systems, such as the genetic feature learning framework by Jiang et al. [37], accurately identify surface irregularities and defects on workpieces. These systems reduce errors caused by lighting sensitivity in single RGB modes. For public safety, multi-target RGB-Depth tracking systems (e.g., the RGBD1K benchmark dataset by Zhu et al. [106]) detect abnormal behaviors and provide early warnings for group events in crowded places like airports and stations through 3D trajectory analysis. Additionally, RGB-Depth tracking's fusion of color and spatial depth cues performs well in structured indoor environments requiring precise 3D localization and occlusion management. In robotic manipulation, it improves object grasping by aligning texture features with depth contours, enhancing performance in cluttered scenes. Healthcare and rehabilitation benefit from detailed skeletal movement capture for patient monitoring and posture correction. Industrial inspection uses RGB-D sensors to track components with similar textures but different height profiles, ensuring strict quality control in manufacturing pipelines. Overall, integrating depth information enhances robustness against illumination changes and provides vital structural disambiguation for tasks needing spatial understanding.

1.3.7 RGB-Thermal Tracking in VOT

Fusing RGB and thermal imaging ensures robust tracking in darkness/smoke by merging color details with heat signatures. Critical for fire rescue (heat-source detection) and military ops (camouflage penetration). This section examines the advantages, limitations, and practical applications of RGB-thermal tracking, sup-

Fig. 1.14 Example of rgb-thermal tracking [42]

ported by recent advancements in tracking frameworks. Figure 1.14 is an example of RGB-Thermal Tracking.

1. Advantages

The multi-modal fusion of RGB and thermal imaging significantly enhances the robustness of target tracking in all-weather environments by leveraging complementary cross-spectral sensing capabilities. Thermal imaging sensors capture the thermal radiation signatures of targets, enabling penetration through visual interferences such as low light, haze, and smoke. This allows continuous tracking of life forms (e.g., trapped individuals during nighttime search and rescue) or high-temperature objects (e.g., vehicle engines and industrial equipment hotspots) [17]. Meanwhile, the RGB modality provides rich texture details—such as clothing colors and license plate information—that assist in distinguishing interfering targets with similar thermal signatures (e.g., differentiating humans from heating devices). A fusion framework based on two-stream convolutional networks [45] maintains tracking continuity during scenarios involving partial occlusion or rapid target motion (e.g., animals fleeing forest fires) by effectively complementing cross-modal features. Compared to single-modal approaches, this method demonstrates greater adaptability in complex thermal environments, such as differentiating sun-heated rocks with temperature profiles close to human bodies [14].

2. Limitations

Cross-modal data alignment faces significant bottlenecks stemming from differences in physical sensor characteristics and limitations in algorithm generalization. The disparity in spatial resolution between thermal imaging and RGB sensors—thermal sensors typically having lower resolution than visible-light cameras—often results in feature matching aberrations, which can cause localization drift, particularly when tracking small targets at long distances [4]. Moreover, the stability of thermal radiation features in dynamic environments is challenged by transient temperature variations, such as when a target moves into an air-conditioned area, leading to failures in traditional temperature-threshold-based segmentation methods [21]. Additionally, cross-modal fusion models demand extremely high synchronization accuracy in hardware; even minor spatiotemporal alignment errors (e.g., sensor mounting biases) can introduce conflicts between modalities. Although online calibration techniques exist—such as the early fusion network proposed by Wang et al. [75]—the balance between real-time performance and calibration accuracy on mobile platforms remains an open challenge.

3. Practical Applications
This technology plays an irreplaceable role in fields such as security surveillance, disaster relief, military reconnaissance, industrial inspection, border patrol, and autonomous driving. In intelligent security systems, thermal-RGB fusion trackers (e.g., the multi-target detection framework proposed by El Ahmar et al. [18]) can accurately distinguish intruders from animals in nighttime or backlit environments, effectively suppress false alarms caused by thermal signals from moving foliage, and maintain continuous monitoring capability from daylight to complete darkness. In autonomous driving, RGB-T systems are employed to sustain vehicle and pedestrian tracking during tunnel transitions, backlit scenes, and night driving, where RGB-only systems tend to fail. Fire rescue robots leverage thermal imaging to detect the temperature signatures of survivors amid fire scenes, combining RGB data to identify obstacle structures and facilitate coordinated search and rescue in complex environments [46]. In military applications, vehicle-mounted multispectral tracking systems enhance persistent target locking on the battlefield by exploiting thermal imaging's anti-camouflage properties alongside RGB-based terrain semantic analysis [17]. In industrial inspection, synergistic thermal-RGB tracking enables early warning of overheating in blast furnace equipment, localizes abnormal heat zones, and analyzes equipment nameplate information via visible light images to support rapid fault diagnosis. Moreover, search-and-rescue UAVs equipped with RGB-T sensors locate heat-emitting survivors in disaster zones with poor or obstructed visual signals. These applications highlight the technology's critical value in safety-sensitive environments subject to variable lighting conditions.

1.3.8 RGB-Language Tracking in VOT

Combining RGB visuals with language semantics enables context-aware tracking using verbal commands (e.g., "Track the red-coated hiker"). This approach bridges visual data and task-specific instructions. It enhances human–robot interaction and improves security systems that require semantic reasoning. This section analyzes the advantages, limitations, and practical applications of RGB-language tracking. Recent advancements in tracking frameworks support these discussions. Figure 1.15 shows examples of RGB-language tracking.

1. Advantages
Multi-modal tracking combining RGB images and natural language greatly improves target description flexibility and tracking robustness in complex scenes. Unlike traditional bounding box or template-based methods, linguistic descriptions

Fig. 1.15 Examples of rgb-language tracking [79]

offer rich semantic details (e.g., "middle-aged man wearing a red plaid shirt and carrying a black backpack"). They help maintain tracking stability despite significant changes in the target's appearance, such as occlusion or clothing changes, by using a text-guided feature relocation mechanism [79]. Methods like the Siamese natural language tracker [24] enable fine-grained differentiation of similar targets (e.g., vehicles of the same model but different colors) by embedding spatial learning and aligning visual features with language semantics in a shared space. In intelligent interaction scenarios, users can dynamically modify the tracking target via real-time voice commands, enabling active perception through human–machine collaboration [90].

2. Limitations

Cross-modal semantic differences and high annotation costs are major challenges for this technology. Natural language ambiguity, such as vague references like "left object" in crowded scenes, often causes unclear tracking targets [51]. Additionally, the gap between visual and linguistic modalities limits models' ability to understand complex syntax, including negations and metaphors. Regarding real-time performance, linguistic encoders like BERT introduce higher computational latency, which can reduce responsiveness, especially on resource-constrained edge devices [23]. Furthermore, the lack of large, high-quality datasets that pair language with video—requiring manual text annotation and temporal alignment—restricts models' generalization to rare cases, such as dialectal expressions and specialized terms [30].

3. Practical Applications

In intelligent security, this technology allows police to perform cross-camera target re-recognition using natural language descriptions (e.g., "wearing a black duck-tongued hat and carrying a silver suitcase") combined with cross-modal retrieval frameworks [50]. In elderly assistive robotics, robots can continuously locate dynamic objects based on voice commands (e.g., "Track my water cup") to help users with limited mobility. In industrial quality inspection, operators describe defect features with specialized terminology (e.g., "track the circular scratch on the right side of the glass surface") to guide vision systems in accurately identifying micron-level defects. The fusion of RGB and natural language offers a powerful paradigm for semantic-level tracking across diverse applications. In human–computer interaction, users refer to objects using spoken or written cues (e.g., "track the red suitcase next to the door"), improving accessibility and assistive technologies. In video surveillance, language-guided tracking enables operators to search for and follow specific entities described in natural language, enhancing operational efficiency. In robotics, instruction-following agents use RGB-language fusion to navigate and manipulate environments based on verbal commands. Additionally, RGB-language tracking supports video content analysis by mapping descriptive phrases to visual targets for summarization or retrieval. Current trends focus on zero-shot language adaptation (e.g., cue learning [79]) and lightweight multi-modal fusion architectures to promote widespread use in applications like augmented reality navigation and intelligent guidance.

1.4 Summary

Visual Object Tracking (VOT) is a fundamental task in computer vision that aims to continuously estimate the state of a target object across video frames based on its initial appearance. This chapter presents an overview of VOT, focusing on its core principles, key challenges, and classification of data modalities. It outlines the main objectives of visual tracking and highlights common difficulties such as occlusion, appearance variation, motion blur, and background clutter. Technologically, VOT has evolved from early correlation filter-based methods to deep learning–driven end-to-end frameworks, significantly enhancing both accuracy and robustness. In parallel, the sensing modalities employed in tracking have expanded beyond traditional RGB and thermal imaging to include depth sensors, LiDAR, and event cameras. Modern tracking systems increasingly rely on multi-modal fusion strategies—such as RGB–LiDAR and RGB–Depth—to improve perception capabilities under challenging conditions and enhance environmental adaptability. By analyzing the theoretical foundations, prevailing challenges, and cross-modal tracking methodologies in VOT, this chapter provides essential insights for developing more resilient and context-aware visual tracking systems.

References

1. Alismail H, Browning B, Lucey S (2016) Robust tracking in low light and sudden illumination changes. In: 2016 Fourth international conference on 3D vision (3DV). IEEE, New York, pp 389–398
2. Asvadi A, et al (2016) 3D object tracking using RGB and LIDAR data. In: 2016 IEEE 19th international conference on intelligent transportation systems (ITSC). IEEE, New York, pp 1255–1260
3. Bay H, Tuytelaars T, Van Gool L (2006) Surf: speeded up robust features. In: Proceedings of the Computer Vision–ECCV 2006: 9th European conference on computer vision, Graz, Austria, May 7–13, 2006, Part I 9. Springer, Berlin, pp 404–417
4. Berg A, Ahlberg J, Felsberg M (2015) A thermal object tracking benchmark. In: 2015 12th IEEE international conference on advanced video and signal based surveillance (AVSS). IEEE, New York, pp 1–6

5. Bertinetto L, et al (2016) Fully-convolutional siamese networks for object tracking. In: Proceedings of the computer vision–ECCV 2016 workshops: Amsterdam, The Netherlands, October 8–10 and 15–16, 2016, part II 14. Springer, Berlin, pp 850–865
6. Cai Y, et al (2023) Robust object modeling for visual tracking. In: Proceedings of the IEEE/CVF international conference on computer vision, pp 9589–9600
7. Camplani M, et al (2017) Multiple human tracking in RGB-depth data: a survey. IET Comput Vision 11(4):265–285
8. Cavallaro A, Steiger O, Ebrahimi T (2005) Tracking video objects in cluttered background. IEEE Trans Circuits Syst Video Technol 15(4):575–584
9. Chen H, et al (2019) Asynchronous tracking-by-detection on adaptive time surfaces for event-based object tracking. In: Proceedings of the 27th ACM international conference on multimedia, pp 473–481
10. Chen R, et al (2022) GFSNet: generalization-friendly siamese network for thermal infrared object tracking. Infrared Phys Technol 123:104190
11. Chen X, et al (2023) Seqtrack: Sequence to sequence learning for visual object tracking. In: Proceedings of the IEEE/CVF conference on computer vision and pattern recognition, pp 14572–14581
12. ColorHub (2023) Environmental pollution: the metaphor of plastic bags and tree branches. https://colorhub.me/photos/gBaDZ
13. Comaniciu D, Ramesh V, Meer P (2003) Kernel-based object tracking. IEEE Trans Pattern Anal Mach Intell 25(5):564–577
14. Cong R, et al (2022) Does thermal really always matter for RGB-T salient object detection? IEEE Trans Multimedia 25:6971–6982
15. Cui Y, et al (2023) Mixformerv2: efficient fully transformer tracking. Adv Neural Inf Proces Syst 36:58736–58751
16. Danaci KI, Akagunduz E (2024) A survey on infrared image & video sets. Multimed Tools Appl 83(6):16485–16523
17. Ahmar WE (2023) Thermal-RGB sensory data for reliable and robust perception. PhD thesis. Université d'Ottawa/University of Ottawa, Ottawa
18. Ahmar WAE, et al (2022) Multiple object detection and tracking in the thermal spectrum. In: Proceedings of the IEEE/CVF conference on computer vision and pattern recognition, pp 277–285
19. Erçelik E, Yurtsever E, Knoll A (2021) 3D object detection with multi-frame RGB-lidar feature alignment. IEEE Access 9:143138–143149
20. Fan B, et al (2024) EasyTrack: efficient and compact one-stream 3D point clouds tracker. arXiv preprint arXiv:2404.05960
21. Felsberg M, et al (2015) The thermal infrared visual object tracking VOT-TIR2015 challenge results. In: Proceedings of the ieee international conference on computer vision workshops, pp 76–88
22. Feng F, Shen B, Liu H (2018) Visual object tracking: in the simultaneous presence of scale variation and occlusion. Syst Sci Control Eng 6(1):456–466
23. Feng Q, et al (2020) Real-time visual object tracking with natural language description. In: Proceedings of the IEEE/CVF winter conference on applications of computer vision, pp 700–709
24. Feng Q, et al (2021) Siamese natural language tracker: tracking by natural language descriptions with Siamese trackers. In: Proceedings of the IEEE/CVF conference on computer vision and pattern recognition, pp 5851–5860
25. Fu Y, et al (2023) Distractor-aware event-based tracking. IEEE Trans Image Process 32:6129–6141
26. Gallego G, et al (2020) Event-based vision: a survey. IEEE Trans Pattern Anal Mach Intell 44(1):154–180
27. Gao L, et al (2025) RGB-D visual object tracking with transformer-based multimodal feature fusion. Knowl-Based Syst 322:113531

28. Gao S, Zhou C, Zhang J (2023) Generalized relation modeling for transformer tracking. In: Proceedings of the IEEE/CVF conference on computer vision and pattern recognition, pp 18686–18695
29. Gopal GY, Amer MA (2024) Separable self and mixed attention transformers for efficient object tracking. In: Proceedings of the IEEE/CVF winter conference on applications of computer vision, pp 6708–6717
30. Guo M, et al (2022) Divert more attention to vision-language tracking. Adv Neural Inf Proces Syst 35:4446–4460
31. Guo Q, et al (2021) Exploring the effects of blur and deblurring to visual object tracking. IEEE Trans Image Process 30:1812–1824. https://doi.org/10.1109/TIP.2020.3045630
32. Guo Q, et al (2021) Exploring the effects of blur and deblurring to visual object tracking. IEEE Trans Image Process 30:1812–1824
33. Guo Q, et al (2021) Learning to adversarially blur visual object tracking. In: Proceedings of the IEEE/CVF international conference on computer vision, pp 10839–10848
34. He X, Zhao L, Chen CY-C (2023) Variable scale learning for visual object tracking. J Ambient Intell Humaniz Comput 14(4):3315–3330
35. Hwang S, et al (2016) Fast multiple objects detection and tracking fusing color camera and 3D LIDAR for intelligent vehicles. In: 2016 13th international conference on ubiquitous robots and ambient intelligence (URAI). IEEE, New York, pp 234–239
36. Ji M, et al (2023) SCTN: event-based object tracking with energy-efficient deep convolutional spiking neural networks. Front Neurosci 17:1123698
37. Jiang M-X, et al (2019) Visual object tracking in RGB-D data via genetic feature learning. Complexity 2019(1):4539410
38. Jiao Y, et al (2024) GSOT3D: towards generic 3D single object tracking in the wild. arXiv preprint arXiv:2412.02129
39. Kang B, et al (2023) Exploring lightweight hierarchical vision transformers for efficient visual tracking. In: Proceedings of the IEEE/CVF international conference on computer vision, pp 9612–9621
40. Kass M, Witkin A, Terzopoulos D (1988) Snakes: active contour models. Int J Comput Vis 1(4):321–331
41. Kiani Galoogahi H, Fagg A, Lucey S (2017) Learning background-aware correlation filters for visual tracking. In: Proceedings of the IEEE international conference on computer vision, pp 1135–1143
42. Kristan M, et al (2020) The eighth visual object tracking VOT2020 challenge results. In: Proceedings of the Computer Vision–ECCV 2020 workshops: Glasgow, UK, August 23–28, 2020, Part V 16. Springer, Berlin, pp 547–601
43. Lee S-H, et al (2022) Markerless 3D skeleton tracking algorithm by merging multiple inaccurate skeleton data from multiple RGB-d sensors. Sensors 22(9):3155
44. Li B, et al (2018) High performance visual tracking with Siamese region proposal network. In: Proceedings of the IEEE conference on computer vision and pattern recognition, pp 8971–8980
45. Li C, et al (2018) Fusing two-stream convolutional neural networks for RGB-T object tracking. Neurocomputing 281:78–85
46. Li C, et al (2019) RGB-T object tracking: Benchmark and baseline. Pattern Recogn 96:106977
47. Li C, et al (2018) RGB-T Object Tracking:Benchmark and Baseline. arXiv: 1805.08982 [cs.CV]. https://arxiv.org/abs/1805.08982
48. Li F, et al (2018) Learning spatial-temporal regularized correlation filters for visual tracking. In: Proceedings of the IEEE conference on computer vision and pattern recognition, pp 4904–4913
49. Li S, et al (2024) Two-stage spatio-temporal feature correlation network for infrared ground target tracking. IEEE Trans Geosci Remote Sens 62:1–14
50. Li Y, et al (2022) Cross-modal target retrieval for tracking by natural language. In: Proceedings of the IEEE/CVF conference on computer vision and pattern recognition, pp 4931–4940

51. Li Z, et al (2017) Tracking by natural language specification. In: Proceedings of the IEEE conference on computer vision and pattern recognition, pp 6495–6503
52. Liu J, et al (2024) M3SOT: multi-frame, multi-field, multi-space 3D single object tracking. In: Proceedings of the AAAI conference on artificial intelligence, vol. 38(4), pp 3630–3638
53. Liu Q, et al (2022) Learning dual-level deep representation for thermal infrared tracking. IEEE Trans Multimedia 25:1269–1281
54. Liu S, et al (2021) Effective template update mechanism in visual tracking with background clutter. Neurocomputing 458:615–625
55. Lowe DG (1999) Object recognition from local scale-invariant features. In: Proceedings of the seventh IEEE international conference on computer vision, vol 2. IEEE, New York, pp 1150–1157
56. Lu Y, et al (2024) VoxelTrack: exploring Voxel Representation for 3D Point Cloud Object Tracking. arXiv preprint arXiv:2408.02263
57. Ma B, et al (2016) Visual tracking under motion blur. IEEE Trans Image Process 25(12):5867–5876
58. Meshgi K, Ishii S (2015) The state-of-the-art in handling occlusions for visual object tracking. IEICE Trans Inform Syst 98(7):1260–1274
59. Muller N, et al (2021) Seeing behind objects for 3d multi-object tracking in rgb-d sequences. In: Proceedings of the IEEE/CVF conference on computer vision and pattern recognition, pp 6071–6080
60. Nguyen HT, Smeulders AWM (2004) Fast occluded object tracking by a robust appearance filter. IEEE Trans Pattern Anal Mach Intell 26(8):1099–1104
61. Nie J, et al (2023) Glt-t: global-local transformer voting for 3d single object tracking in point clouds. In: proceedings of the AAAI conference on artificial intelligence, vol 37(2):1957–1965
62. Nie J, et al (2024) Towards category unification of 3d single object tracking on point clouds. arXiv preprint arXiv:2401.11204
63. Nilsback M-E, Zisserman A (2008) Automated flower classification over a large number of classes. In: 2008 Sixth Indian conference on computer vision, graphics & image processing. IEEE, New York, pp 722–729
64. Pan J, Hu B (2007) Robust occlusion handling in object tracking. In: 2007 IEEE conference on computer vision and pattern recognition. IEEE, New York, pp 1–8
65. Panda KD, Meher S (2011) Robust real-time object tracking under background clutter. In 2011 International conference on image information processing. IEEE, New York, pp 1–6
66. Peng Z, et al (2023) 3-D objects detection and tracking using solid-state Li- DAR and RGB camera. IEEE Sensors J 23(13):14795–14808
67. Qi H, et al (2020) P2b: point-to-box network for 3d object tracking in point clouds. In: Proceedings of the IEEE/CVF conference on computer vision and pattern recognition, pp 6329–6338
68. Qi M, et al (2024) Exploring reliable infrared object tracking with spatio-temporal fusion transformer. Knowl-Based Syst 284:111234
69. Samal K, et al (2021) Task-driven rgb-lidar fusion for object tracking in resource-efficient autonomous system. IEEE Trans Intell Veh 7(1):102–112
70. Seo B-K, et al (2013) Optimal local searching for fast and robust textureless 3D object tracking in highly cluttered backgrounds. IEEE Trans Vis Comput Graph 20(1):99–110
71. Shan J, et al (2021) PTT: point-track-transformer module for 3D single object tracking in point clouds. In: 2021 IEEE/RSJ international conference on intelligent robots and systems (IROS). IEEE, New York, pp 1310–1316
72. Shan Y, et al (2025) SDTrack: a baseline for event-based tracking via spiking neural networks. arXiv preprint arXiv:2503.08703
73. Shi L, et al (2024) Explicit visual prompts for visual object tracking. In: Proceedings of the AAAI conference on artificial intelligence, vol. 38(5), pp 4838–4846
74. Sualeh M, Kim G-W (2020) Visual-LiDAR based 3D object detection and tracking for embedded systems. IEEE Access 8:156285–156298

75. Wang H, et al (2023) Thermal images-aware guided early fusion network for cross-illumination RGB-T salient object detection. Eng Appl Artif Intell 118:105640
76. Wang H, et al (2019) Multi-3D-object tracking by fusing RGB and 3D-LiDAR data. In: 2019 IEEE international conference on unmanned systems (ICUS). IEEE, New York, pp 941–946
77. Wang S, et al (2025) Event stream-based visual object tracking: HDETrack V2 and a high-definition benchmark. arXiv preprint arXiv:2502.05574
78. Wang X, et al (2024) Mambaevt: event stream based visual object tracking using state space model. arXiv preprint arXiv:2408.10487
79. Wang X, et al (2021) Towards more flexible and accurate object tracking with natural language: algorithms and benchmark. In: Proceedings of the IEEE/CVF conference on computer vision and pattern recognition, pp 13763–13773
80. Wu Q, et al (2024) Boosting 3D single object tracking with 2D matching distillation and 3D Pre-training. In: European conference on computer vision. Springer, Berlin, pp 270–288
81. Wu Q, et al (2023) Dropmae: masked autoencoders with spatial-attention dropout for tracking tasks. In: Proceedings of the IEEE/CVF conference on computer vision and pattern recognition, pp 14561–14571
82. Wu S, et al (2021) Aircraft tracking in infrared imagery with adaptive learning and interference suppression. Electron Lett 57(16):636–638
83. Xiao J, et al (2017) Robust fusion of color and depth data for RGB-D target tracking using adaptive range-invariant depth models and spatio-temporal consistency constraints. IEEE Trans Cybern 48(8):2485–2499
84. Xiao Y, et al (2022) Attribute-based progressive fusion network for rgbt tracking. In: Proceedings of the AAAI conference on artificial intelligence, vol. 36(3), pp 2831–2838
85. Xie F, et al (2023) Videotrack: learning to track objects via video transformer. In: Proceedings of the IEEE/CVF conference on computer vision and pattern recognition, pp 22826–22835
86. Xu C, et al (2017) Lie-x: depth image based articulated object pose estimation, tracking, and action recognition on lie groups. Int J Comput Vis 123:454–478
87. Xu T-X, et al (2023) Mbptrack: improving 3d point cloud tracking with memory networks and box priors. In: Proceedings of the IEEE/CVF international conference on computer vision, pp 9911–9920
88. Yang J, et al (2022) Rgbd object tracking: an in-depth review. arXiv preprint arXiv:2203.14134
89. Yang Y, et al (2024) SiamMo: Siamese Motion-Centric 3D object tracking. arXiv preprint arXiv:2408.01688
90. Yang Z, et al (2020) Grounding-tracking-integration. IEEE Trans Circuits Syst Video Technol 31(9):3433–3443
91. Yi A, Anantrasirichai N (2024) A comprehensive study of object tracking in low-light environments. Sensors 24(13):4359
92. Yuan D, et al (2022) Structural target-aware model for thermal infrared tracking. Neurocomputing 491:44–56
93. Zhang B, et al (2024) SVR-AVT: scale variation robust active visual tracking. In 2024 International joint conference on neural networks (IJCNN), pp 1–8. https://doi.org/10.1109/IJCNN60899.2024.10650947
94. Zhang H, Yin Z, Zhang H (2023) Thermal infrared object tracking using correlation filters improved by level set. Signal, Image Video Process 17(3):791–797
95. Zhang J, et al (2024) Robust 3D tracking with quality-aware shape completion. In: Proceedings of the AAAI conference on artificial intelligence, vol. 38(7):7160–7168
96. Zhang S, et al (2022) Evtracker: an event-driven spatiotemporal method for dynamic object tracking. Sensors 22(16):6090
97. Zhao H, Wang D, Lu H (2023) Representation learning for visual object tracking by masked appearance transfer. In: Proceedings of the IEEE/CVF conference on computer vision and pattern recognition, pp 18696–18705

98. Zheng C, et al (2022) Beyond 3d siamese tracking: a motion-centric paradigm for 3d single object tracking in point clouds. In: Proceedings of the IEEE/CVF conference on computer vision and pattern recognition, pp 8111–8120
99. Zheng C, et al (2021) Box-aware feature enhancement for single object tracking on point clouds. In: Proceedings of the IEEE/CVF international conference on computer vision, pp 13199–13208
100. Zheng X, et al (2023) Deep learning for event-based vision: a comprehensive survey and benchmarks. arXiv preprint arXiv:2302.08890
101. Zheng Y, et al (2024) Odtrack: online dense temporal token learning for visual tracking. In: Proceedings of the AAAI conference on artificial intelligence, vol. 38(7), pp 7588–7596
102. Zhong P, et al (2024) Low-light object tracking: a benchmark. arXiv preprint arXiv:2408.11463
103. Zhou C, et al (2022) PTTR: relational 3d point cloud object tracking with transformer. In: Proceedings of the IEEE/CVF conference on computer vision and pattern recognition, pp 8531–8540
104. Zhou X, et al (2023) Reading relevant feature from global representation memory for visual object tracking. Adv Neural Inf Proces Syst 36:10814–10827
105. Zhu X-F, Xu T, Wu X-J (2022) Visual object tracking on multi-modal RGB-D videos: A review. arXiv preprint arXiv:2201.09207
106. Zhu X-F, et al (2023) RGBD1K: a large-scale dataset and benchmark for RGB-D object tracking. In: Proceedings of the AAAI conference on artificial intelligence, vol. 37(3):3870–3878

Chapter 2
Fundamental Knowledge and Frameworks

Abstract This chapter provides a comprehensive overview of the foundational knowledge and frameworks in visual object tracking (VOT). It begins with an exploration of key deep learning models—Convolutional Neural Networks (CNNs), Multi-layer Perceptrons (MLPs), and Transformers—highlighting their structural designs, learning mechanisms, and roles in visual representation and temporal modeling. The chapter then introduces mainstream tracking paradigms including Siamese Networks, Discriminative Correlation Filters (DCFs), and One-Stream Frameworks, explaining their architectures, interaction strategies, and optimization objectives. These paradigms are discussed in the context of their adaptability to real-time and complex scenarios, such as occlusion, scale variation, and background interference. The evolution from handcrafted to learned features and from decoupled to unified modeling structures is emphasized. The chapter concludes by addressing current challenges and outlining future research directions. This foundational understanding supports readers in grasping advanced tracking algorithms and sets the stage for domain-specific discussions in later chapters.

2.1 Deep Learning Foundations

2.1.1 Convolutional Neural Network

Convolutional Neural Networks (CNNs) are core architectures in deep learning and are widely applied in computer vision, primarily for tasks such as image classification, object detection, and image segmentation. In visual object tracking (VOT) tasks, CNNs achieve continuous and precise localization of specific targets through dynamic feature modeling and spatiotemporal correlation learning. The core architecture of CNNs comprises convolutional layers (extracting local features), pooling layers (reducing dimensionality while retaining key information), activation functions (introducing nonlinearity), and fully connected layers (facilitating high-level reasoning) [30, 45]. Compared to traditional neural networks, CNNs significantly reduce computational costs through the use of local receptive

M. Wang et al., *Visual Object Tracking across Modalities*, Advances in Computer Vision and Pattern Recognition, https://doi.org/10.1007/978-981-95-3664-1_2

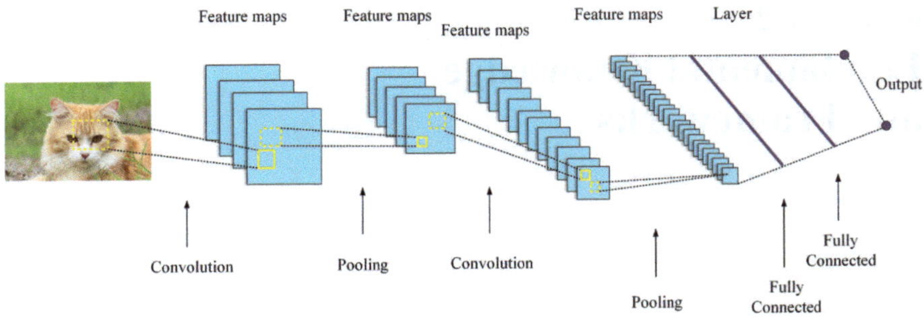

Fig. 2.1 Schematic diagram of convolutional neural network architecture

fields and parameter sharing, leveraging pooling operations to enhance the model's invariance to translation and scaling. These characteristics have established CNNs as a cornerstone of computer vision [45].

Convolutional neural networks have demonstrated remarkable performance in handling complex data tasks, owing to their unique hierarchical architecture. Firstly, through an end-to-end learning mechanism, CNNs enable automatic feature extraction from low-level visual features to high-level semantic concepts, thereby transforming the traditional paradigm that relies on manually designed feature extractors. This has led to significant breakthroughs in various challenging visual tasks. Secondly, CNNs exhibit strong translation invariance and are designed with local receptive fields, which allow them to robustly process spatially correlated data such as images, videos, and language, ensuring reliable recognition despite variations in object positions. Furthermore, by leveraging local connectivity and weight sharing, CNNs significantly reduce model complexity while enhancing generalization capability.

CNNs leverage their powerful hierarchical feature extraction capability (e.g., through backbone networks like ResNet [34] and MobileNet [39]) to extract the appearance features (such as shape, color, and motion patterns) of a given target from each frame in a video sequence. Using similarity matching (e.g., via Siamese network architectures), they then search for regions associated with the initial target in subsequent frames. Additionally, by incorporating online update mechanisms such as template adaptation or memory networks, CNNs can adapt to appearance changes caused by occlusion or deformation, as well as interference from complex backgrounds.

The typical architecture of CNNs consists of an input layer, stacked convolutional layers, pooling layers, fully connected layers, and an output layer, with each layer working collaboratively to enable efficient feature learning and decision-making. The architecture of the convolutional neural network is illustrated in the Fig. 2.1.

Input Layer In CNN architectures, the input layer constitutes the network's entry point, tasked with ingesting and preprocessing raw input data (e.g., images, audio signals). In computer vision applications, this layer typically standardizes input values and resizes images to fixed dimensions, thereby facilitating efficient processing by downstream convolutional layers. The input data dimensionality is conventionally specified in terms of height, width, and channel depth. The input layer may additionally implement data augmentation techniques (including rotation and random cropping) to improve model generalization. Following preprocessing, the data undergo immediate transformation through the initial convolutional kernel. This layer's fundamental role involves converting raw input signals into high-dimensional learnable representations that facilitate efficient hierarchical pattern recognition by subsequent network layers, thereby establishing the essential groundwork for effective feature extraction. The input layer receives the original image, which is usually represented as a three-dimensional tensor:

$$\mathbf{X} \in \mathbb{R}^{H \times W \times C} \tag{2.1}$$

where H is the height of the image, W is the width of the image, C is the number of channels, e.g.. $C = 3$ for an RGB image.

Convolutional Layer Convolutional layers are the core of convolutional neural networks. They extract local features such as edges and textures by sliding convolutional kernels (filters) over the input data. Each kernel computes the dot product of a local region, generating feature maps that capture the spatial distribution of specific patterns in the input. The convolution operation captures local patterns through regional connectivity, significantly reducing the number of parameters by using weight sharing while preserving the translational invariance of the data. The output feature map of a convolutional layer can be formulated as:

$$Y_{i,j}^{(k)} = \sum_{m=1}^{M} \sum_{u=1}^{h} \sum_{v=1}^{w} K_{u,v}^{(k,m)} \cdot X_{i+u,\, j+v}^{(m)} + b^{(k)} \tag{2.2}$$

where $X \in \mathbb{R}^{H \times W \times M}$ is the input feature map with M channels, $K^{(k,m)} \in \mathbb{R}^{h \times w}$ is the convolutional kernel connecting input channel m to output channel k, $b^{(k)}$ is the bias term for the k-th output channel, $Y_{i,j}^{(k)}$ is the value at position (i, j) in the k-th output feature map, h, w are the height and width of the convolution kernel.

Pooling Layer A pooling layer is used to reduce the spatial dimensions of feature maps, thereby decreasing computational complexity and enhancing translational invariance. Common operations include max pooling, which preserves prominent features by selecting the maximum value within a local region, and average pooling, which smooths features by computing the mean value within a local area. Other pooling variants have been proposed to improve generalization or retain more information. By discarding less relevant information while preserving key features, pooling helps prevent overfitting and progressively enlarges the receptive field, enabling higher-layer representations to capture more global semantic information. Although pooling may lead to the loss of some spatial details, its efficient down-sampling capability makes it an indispensable component of convolutional neural networks (CNNs). The most common operation is maximum pooling, the output of a max pooling layer can be expressed as:

$$Y_{i,j}^{(k)} = \max_{(u,v)\in\Omega} X_{s\cdot i+u,\, s\cdot j+v}^{(k)} \tag{2.3}$$

where $X \in \mathbb{R}^{H\times W\times C}$ is the input feature map, $Y \in \mathbb{R}^{H'\times W'\times C}$ is the output feature map, (i, j) is the spatial location in the output, k is the channel index (same as input), Ω is the pooling window, and s is the stride (pooling step size).

Fully Connected Layer The fully connected layer is typically located at the end of a neural network and serves to map high-level feature representations to the output space. Each neuron in this layer is connected to all neurons in the preceding layer, enabling the integration of global semantic information. Due to the large number of parameters, fully connected layers are prone to overfitting, which is often mitigated through techniques such as dropout or weight regularization. In modern convolutional neural networks (CNNs), fully connected layers are increasingly replaced by global average pooling (GAP) to reduce computational complexity and improve model generalization. The output of the fully connected layer can be defined as:

$$\mathbf{y} = \sigma(\mathbf{W}\mathbf{x} + \mathbf{b}) \tag{2.4}$$

where $\mathbf{x} \in \mathbb{R}^n$ is the input vector, $\mathbf{W} \in \mathbb{R}^{m\times n}$ is the weight matrix, $\mathbf{b} \in \mathbb{R}^m$ is the bias vector, $\sigma(\cdot)$ is the activation function (e.g., ReLU [29], Sigmoid), and $\mathbf{y} \in \mathbb{R}^m$ is the output vector.

Output Layer The output layer is the last layer of the whole network, which is responsible for making the final prediction results according to the features learned by the network. Different tasks have different output requirements, and the structure and activation function of the output layer will change accordingly. In classification tasks, the output layer typically applies a softmax function to produce class probabilities:

$$\hat{y}_i = \frac{e^{z_i}}{\sum_{j=1}^{C} e^{z_j}}, \quad \text{for } i = 1, 2, \ldots, C \tag{2.5}$$

where $\mathbf{z} = \mathbf{W}\mathbf{x} + \mathbf{b} \in \mathbb{R}^C$ is the logits (raw output before softmax), C is the number of classes, and $\hat{y}_i$ is the predicted probability of class i, $\sum_{i=1}^{C} \hat{y}_i = 1$ means output is a valid probability distribution.

Improvements in Convolutional Neural Networks Numerous researchers have significantly enhanced the performance of convolutional neural networks (CNNs) by refining convolution designs, optimizing pooling strategies, and introducing novel activation and loss functions. At the same time, techniques such as acceleration based on Fast Fourier Transform (FFT) [4], low-precision computation, and weight compression [30] have substantially improved the computational efficiency of CNN models. Subsequent researchers further proposed the Winograd minimal filtering algorithm, a class of fast convolutional algorithms designed for small filters (e.g., 3×3) [42].

In VOT, Convolutional Neural Networks (CNNs) play a central role by serving as powerful feature extractors and matching modules, which enable robust and accurate tracking under various challenging conditions, such as illumination changes, occlusion, deformation, and background clutter. At the core of most modern tracking algorithms, CNNs are used to extract high-level semantic features from both the target template and the current search region. Compared to traditional hand-crafted features, CNN-based features are more robust, discriminative, and invariant to changes in appearance. This makes them ideal for representing the visual characteristics of the object being tracked.

2.1.2 Multi-Layer Perceptron Network

Multi-layer perceptron (MLP) is a fundamental feedforward neural network architecture. In object tracking, MLPs primarily function as feature fusion and decision-making modules [50]. Their fully connected nature allows for the effective integration of multi-level features from convolutional neural networks (CNNs) or other feature extraction backbones, enabling nonlinear combinations of local appearance features and global contextual information. By learning complex mappings between the target template and the search region, MLPs can effectively handle challenges such as target deformation and occlusion in object tracking.

Multi-layer perceptron (MLP) has significant advantages in object tracking. Firstly, their strong nonlinear fitting capability allows them to model the complex decision boundaries between the target appearance and the background, improving tracking accuracy. Secondly, the parameter-sharing mechanism in MLP provides a certain degree of invariance to scale and pose variations of the input target, enhancing tracking robustness. Finally, compared to pure convolutional networks, MLPs are more flexible in fusing spatiotemporal features and exhibit better memory capabilities when handling long-term tracking tasks.

Basic Structure A multi-layer perceptron (MLP) is composed of multiple stacked fully connected layers. Its core structure includes an input layer, hidden layers, and an output layer, with nonlinear activation functions enhancing the model's expressiveness. The input layer is responsible for receiving raw data; the hidden layers perform nonlinear transformations of the input through fully connected operations; and the output layer generates the final prediction. This hierarchical design facilitates the forward propagation of data through weight matrices and bias vectors and utilizes backpropagation to optimize parameters, thereby addressing the nonlinear issues that a single-layer perceptron cannot handle.

The formula for the MLP can be expressed as follows:

$$z^{(l)} = W^{(l)} a^{(l-1)} + b^{(l)} \tag{2.6}$$

$$a^{(l)} = f\left(z^{(l)}\right) \tag{2.7}$$

where $W^{(l)}$ is the weight matrix (connecting layer $l - 1$ and layer l), $b^{(l)}$ is the bias vector, $a^{(l-1)}$ is the output of the previous layer (with the input layer defined as $a^{(0)} = x$), and $f(\cdot)$ denotes the activation function (e.g., ReLU [29], Sigmoid).

The key component of MLPs lies in nonlinear activation functions, which enable the model to approximate complex data patterns. An activation function is a nonlinear transformation applied within a neural network, typically imposed after the weighted summation of a neuron's inputs. Its mathematical formulation is commonly expressed as:

$$y = f(z), \quad z = \sum W_i \cdot x_i + b \tag{2.8}$$

where z is the weighted sum of inputs, w_i represents the weight, x_i denotes the input, b is the bias term, and $f(\cdot)$ is the activation function.

Activation functions play a critical role in neural networks by introducing nonlinearity, determining neuron outputs, influencing gradient propagation, and controlling the output range. As a core component of neural networks, activation functions directly affect the model's nonlinear representational capacity, training stability, and overall performance. Based on the above analysis and comprehensive considerations, selecting appropriate activation functions for each layer of an MLP is essential for model optimization. A well-chosen activation function can enhance the expressive power of the network and help mitigate issues such as vanishing or exploding gradients.

Tanh The hyperbolic tangent function is a commonly used nonlinear activation function with an output range between -1 and 1. It is zero-centered, which helps alleviate the vanishing gradient problem and accelerates training convergence. Tanh is suitable for scenarios where both positive and negative activations are required. Although it may still suffer from gradient saturation in deep neural networks, it remains applicable in certain sequence models such as RNNs and LSTMs [50]. The function is defined as follows:

$$\tanh(x) = \frac{e^x - e^{-x}}{e^x + e^{-x}} \tag{2.9}$$

Output values in the range of (–1, 1) provide nonlinearity, making it suitable for modeling a wide variety of data distributions. Although the tanh function helps mitigate the vanishing gradient problem compared to sigmoid, its outputs are bounded, which can still lead to gradient saturation for large input values, thereby slowing down learning in deep networks.

Sigmoid The sigmoid function is a classic S-shaped activation function with an output range between 0 and 1. It smoothly maps input values into a probabilistic space, making it suitable for probability-based outputs. However, the sigmoid function suffers from issues such as gradient saturation and non-zero-centered outputs. As a result, it has been gradually replaced by more efficient activation functions like ReLU [29] in modern deep learning. Nevertheless, it remains useful in specific scenarios, such as probabilistic output tasks or in the gating mechanisms of LSTM networks [50]. The formula of the function is given as follows:

$$\sigma(x) = \frac{1}{1 + e^{-x}} \tag{2.10}$$

The S-shaped curve maps real-valued inputs to a range between 0 and 1, making it useful for probability estimation and binary classification tasks. Despite its widespread use in early neural networks, the sigmoid function suffers from the vanishing gradient problem for large positive or negative input values, which can hinder the training of deep networks by significantly slowing down gradient-based optimization.

ReLU [29] The Rectified Linear Unit is a widely used activation function [1], commonly employed in the hidden layers of neural networks. It features nonlinearity, sparse activation, and computational efficiency. However, ReLU [29] suffers from the so-called "dying ReLU" problem, where certain neurons may permanently cease activation. Subsequent variants, such as PReLU [35], have been proposed to address this issue. The function is defined as follows:

$$\text{ReLU}(x) = \max(0, x) \tag{2.11}$$

ReLU [29] helps mitigate the vanishing gradient problem by allowing gradients to flow freely for positive input values, making it particularly effective for deep neural networks, although it can suffer from the "dying ReLU" problem when neurons output only zeros for all inputs.

2.1.3 Transformer

Transformer is a deep learning framework based on the self-attention mechanism, proposed by Vaswani et al. [55] in 2017, and is primarily used for natural language processing tasks. Transformer [55] is a pre-trained language model, most of which employs self-supervised learning (where the training objectives can be automatically computed based on the model's input) to train on large amounts of raw language data, meaning that, unlike traditional methods, it does not require manually annotated data during training.

The core idea of the Transformer is that it does not rely on traditional recurrent or convolutional structures, but captures dependencies between different positions in the input sequence through the self-attention mechanism. A Transformer consists of an encoder and a decoder, each containing multiple layers of identical structure. Its key components include the multi-head self-attention mechanism (MSA), positional encoding, feed-forward neural networks, and residual connections [55]. The self-attention mechanism dynamically assigns weights to each position by calculating the correlation between the query, key, and value, thereby effectively modeling global information. In addition, the Transformer [55] introduces positional encoding to retain the sequence order information, compensating for the insensitivity of the self-attention mechanism to positions. This architecture, due to its strong parallel computation capability and excellent long-distance dependency modeling ability, has rapidly expanded to fields such as computer vision and speech recognition.

In VOT tasks, the application of Transformers mainly focuses on how to leverage their powerful global modeling ability to enhance the robustness and accuracy of tracking. Traditional tracking methods typically rely on convolutional neural networks (CNNs) to extract local features of the target, but they struggle to effectively handle complex scenarios such as target occlusion, deformation, or rapid motion. In contrast, Transformer-based trackers, through the self-attention mechanism, can directly model the global relationships between the target and the search area and even between different positions within the search area. The decoder structure of the Transformer can also be used for iterative prediction of the target bounding box, gradually refining the tracking results through the interaction between query vectors and feature maps. Such methods have shown significant advantages on benchmark datasets like OTB and LaSOT [27], particularly excelling in long-term tracking and handling complex background interference. However, they also face challenges such as high computational overhead and limited adaptability to small samples. Current research is exploring directions such as lightweight design and spatiotemporal joint modeling to further advance the practical application of the Transformer in VOT.

Transformer Model Structures Shown in Fig. 2.2

- **Input.** The input to the Transformer consists of raw data, positional encoding, and optional segment information. In text-related tasks, the input is first converted into vector representations through a word embedding layer. The word embeddings are then summed with the positional embeddings to obtain the word representation vector x, which serves as the input to the Transformer. This structure enables the Transformer to capture both the semantic information of sequential data and the positional relationships among elements, thereby providing ordered input for subsequent self-attention computations.
- **Self-Attention.** This component is used to receive the input (i.e., the word representation vector x) or the output from the previous encoder block. After applying linear transformations, it produces the query (Q), key (K), and value

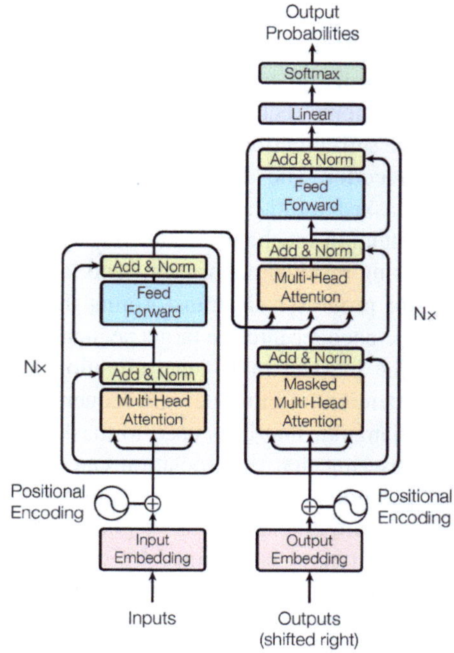

Fig. 2.2 The Transformer-model architecture [55]

(V) matrices. These matrices are then used to compute the output of the self-attention mechanism, whose core formulation is given by:

$$\text{Attention}(Q, K, V) = \text{SoftMax}\left(\frac{QK^\top}{\sqrt{d_k}} + B\right)V \tag{2.12}$$

Here, d_k denotes the number of columns in the Q and K matrices (i.e., the vector dimension). The formula computes the dot product between each row vector of Q and K, scaled by the square root of d_k (a scaling factor) to prevent excessively large inner products that could lead to vanishing gradients. B is an optional relative position bias introduced to enhance spatial awareness. The attention weights are normalized using the *SoftMax* function, and the final output is obtained by computing the weighted sum of V.

Self-Attention captures global dependencies directly through dynamic weight assignment, thereby avoiding the limitations inherent in RNNs and CNNs. It is considered a key factor underlying the powerful representational capacity of the Transformer.

- **Encoder.** The encoder is composed of N identical stacked layers, each consisting of a multi-head self-attention mechanism and a feed-forward neural network (FFN), both optimized through residual connections and layer normalization. The final output is a context-aware representation. The encoder transforms the input into learnable vectors, where the output at each position integrates information from all other positions, thereby providing the decoder with rich semantic and syntactic features [55]. This high-level feature encoding supports various downstream tasks such as machine translation and text classification. Unlike traditional recurrent or convolutional operations, this structure relies entirely on the self-attention mechanism, enabling efficient parallel computation.

- **Decoder.** The decoder consists of multiple stacked layers, each comprising a masked multi-head self-attention mechanism, an encoder–decoder multi-head attention mechanism, and a feed-forward neural network (FFN), all optimized using residual connections and layer normalization. This module is responsible for progressively transforming the encoded source sequence information from the encoder into the target sequence output.
  The masked multi-head attention layer extends the standard multi-head self-attention by introducing a sequence masking mechanism. Specifically, an upper triangular matrix is used to mask future positions beyond the current timestep, ensuring that each position can only attend to tokens at or before the current position. This enforces the autoregressive property of sequence generation. After masking, the attention weights are normalized via the softmax function, and the final output is obtained by computing the weighted aggregation of the value matrix and concatenating the outputs from all attention heads. The decoder architecture can model global dependencies more efficiently than traditional methods, while also avoiding the vanishing gradient problem commonly encountered in RNNs.
- **Multi-Head Attention.** Multi-head attention is a core component of the Transformer architecture, consisting of multiple self-attention layers, designed based on parallelized multi-head attention. The "head" in multi-head attention refers to each attention mechanism, where each head has its own set of weights used to compute the self-attention of the input.
  Operation Process of Multi-Head Attention: Multi-head attention operates by linearly projecting the input sequence's Q, K, and V into multiple distinct self-attention mechanisms. The attention weights are independently computed within each subspace, and the outputs from all heads are then concatenated together. Finally, the concatenated output is passed through a linear layer for transformation and integration to produce the final output. The dimensions of the input and output matrices are consistently maintained.
  The role of Multi-Head Attention is to perform parallel computation of multiple independent attention mechanisms, forming several sub-semantic spaces. This allows the model to focus on information from different dimensions of the semantic space, thereby enhancing its ability to capture local dependencies [55]. This structure follows a divide-and-conquer strategy, enabling the model to learn diverse relationships in different subspaces. Through the collaborative effect of the multi-head mechanism, the model can handle sequences with complex interaction patterns, overcoming the limitations of a single attention mechanism. Multi-head attention improves the computational efficiency of the model through

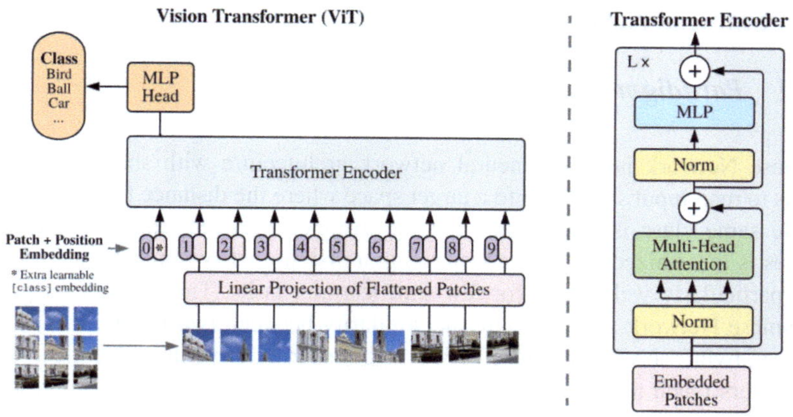

Fig. 2.3 Vision Transformer Network Structure [25]

parallel processing, while the parameter-sharing linear projection layers enhance the model's scalability.

Vision Transformer (ViT) [25] is a model that applies the Transformer architecture to computer vision tasks. Its core idea is to divide the input image into non-overlapping patches of fixed size and flatten these patches into a sequence, which is then used as input for the Transformer. ViT [25] is used primarily for image recognition and computer vision tasks, as shown in Fig. 2.3. The image patches are processed through multiple layers of a Transformer encoder. This encoder consists of self-attention mechanisms and feedforward neural networks, which effectively model global contextual information about the image and capture long-range dependencies between different patches. Ultimately, the model performs task-specific head processing (such as classification or regression) on the feature representations from the encoder's output, enabling efficient handling of various visual tasks. The advantage of this architecture lies in its ability to overcome the limitations of the local receptive field in traditional convolutional neural networks, demonstrating strong feature representation capabilities, especially when trained on large-scale datasets.

The structure of ViT [25] mainly includes Patch Embedding, positional encoding, multiple layers of Transformer Encoder (composed of multi-head self-attention MSA and feed-forward network MLP), and a classification head. It offers global modeling capabilities, a simplified architecture, and scalability. However, its limitations, such as high computational complexity, lack of multi-scale features, and dependency on large-scale pretraining, were addressed in the subsequent Swin Transformer [47] through operations like hierarchical feature pyramids, sliding window self-attention, and relative position bias.

2.2 Siamese Networks

2.2.1 Paradigm Introduction

Siamese Network is a twin neural network architecture with shared weights. It learns to map input samples into a target space where the distance between samples of the same class is minimized, and the distance between samples of different classes is maximized, thereby constructing a discriminative similarity metric [13]. It is particularly well-suited for few-shot learning tasks. The primary purpose of a Siamese Network is to measure the similarity between two inputs. It is referred to as a "twin" network because of its weight-sharing structure—identical weight parameters (such as convolutional kernels and fully connected layer weights) are used in the subnetworks processing the two inputs. As a result, input pairs (e.g., two images or two text segments) are mapped into the same feature space using identical feature extraction logic, ensuring consistency in feature encoding.

Weight sharing plays an irreplaceable role in Siamese Networks. Suppose the weights of the two branches are not shared. In that case, the input pair may be mapped into different feature spaces—for example, one branch might extract shape features while the other extracts color features, rendering the similarity measurement meaningless. Sharing weights effectively reduces the number of network parameters by half, helping to prevent overfitting caused by excessive parameters, and forces the network to learn common features between input pairs, thereby enhancing the model's robustness to input differences. The implementation involves both forward and backward propagation: during forward propagation, the two inputs pass through identical network structures to generate feature vectors; in backward propagation, the losses from both branches are first computed, their gradients are summed, and the shared weight parameters are subsequently updated, ensuring that the weights of both branches remain synchronized.

In certain variants of Siamese Networks applied to object-tracking tasks, the network is explicitly divided into a template branch and a search branch. Although the two branches share the backbone weights, their functional roles are distinct. The template branch takes the initial frame of the target or a reference image as input to extract the target's reference features—essentially "memorizing" the target. It outputs a fixed feature vector or feature map representing the target, which is primarily used for object representation and few-shot classification. In contrast, the search branch processes candidate regions from subsequent frames to extract candidate features from the search area, effectively "searching" for the target. It outputs multi-scale feature maps of the search region, mainly for object tracking and similarity matching. The two branches compute similarity via cross-correlation, producing a response map to localize the target. This dual-branch design enables efficient object tracking or matching: the template branch memorizes the target features, while the search branch performs dynamic searching, achieving a balance between speed and robustness.

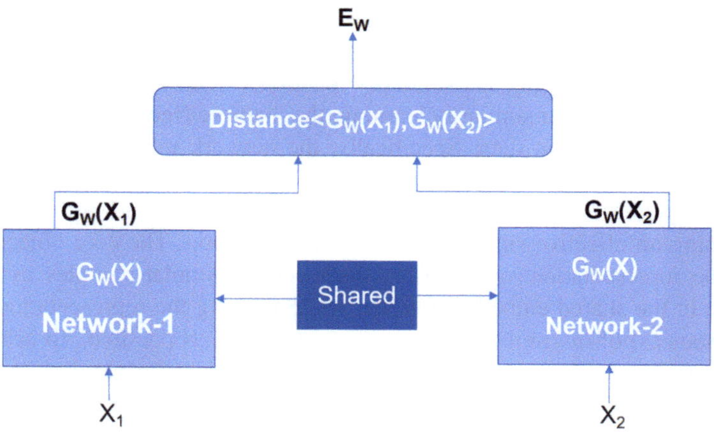

Fig. 2.4 Siamese Network Architecture

The Siamese Network architecture consists of two or more subnetworks that share weights. These subnetworks jointly process input samples by extracting features separately from each input and mapping the original data into a low-dimensional embedding space. The network structure is illustrated in Fig. 2.4:

Shared-Weight Subnetwork (GW) Two inputs are passed through the same neural network architecture, in which each subnetwork is composed of convolutional layers, fully connected layers, etc., to generate feature vectors. The outputs are low-dimensional feature vectors denoted as $GW(X_1)$ and $GW(X_2)$.

Distance Measurement Module This module is used to compute the distance between the two feature vectors. Common approaches include:

- **Euclidean Distance:** Directly measures geometric distance in the feature space.
- **Cosine Similarity:** Computes directional similarity through the inner product of vectors.

The final output is a scalar energy value representing the similarity between the inputs, where 0 indicates dissimilarity and 1 indicates high similarity. The formulas are defined as:

$$D_{\text{Euclidean}}(X_1, X_2) = \|GW(X_1) - GW(X_2)\|_2 \tag{2.13}$$

$$\text{Sim}_{\cos}(X_1, X_2) = \frac{GW(X_1) \cdot GW(X_2)}{\|GW(X_1)\| \, \|GW(X_2)\|} \tag{2.14}$$

The core task of the Siamese Network is to effectively measure the similarity between two input samples by learning their feature representations. It is widely used in similarity matching, object tracking, few-shot classification, anomaly detection, and text-matching tasks. The core task can be broken down into several aspects: similarity measurement, contrastive learning, target matching, and relationship modeling, among others. Similarity measurement is mainly achieved through feature embedding and similarity functions, where similar inputs are closer in the feature space. For similarity learning, the Siamese Network learns the feature representations of input data. Specifically, the network maps each input sample to a feature vector and then computes the distance between feature vectors to measure similarity. The training strategy of the Siamese network primarily focuses on learning an effective similarity measurement method. The core objective is to bring the representation vectors (i.e., distances) of similar samples as close as possible in the shared embedding space, while pushing the representation vectors (i.e., distances) of dissimilar sample pairs as far apart as possible. To achieve this goal, the training process typically uses contrastive loss and triplet loss as the main optimization objective functions, along with positive and negative sample strategies.

In the sample construction process, positive sample pairs generally refer to instances that come from the same category or object, even though they might undergo different acquisition times, sensor transformations, data augmentation, or other variations. Negative sample pairs, on the other hand, are made up of sample instances from different categories or objects. The objective function of the Siamese Network, driven by a contrastive learning mechanism, aims to automatically learn the similarity or dissimilarity relationships between pairs of samples. Specifically, under the condition of shared weights, the network separately extracts the feature representations of input sample pairs, and by optimizing the loss function, it minimizes the feature distance between similar samples while maximizing the feature distance between dissimilar samples, thus forming a well-discriminative feature space. In Siamese Networks, common objective functions include contrastive loss and triplet loss.

The contrastive loss function is based on the label information of paired samples, encouraging positive pairs to be close in the feature space, while ensuring that the distance between negative pairs is at least a predefined margin. The specific formula is as follows:

$$L = \frac{1}{2} y \cdot d^2 + \frac{1}{2}(1 - y) \cdot \max(0, m - d)^2 \tag{2.15}$$

where y is the label, with $y = 0$ indicating a positive sample pair and $y = 1$ indicating a negative sample pair. d is the distance between the two feature vectors, and m is a predefined margin that controls the minimum distance between negative sample pairs. This loss function penalizes the cases where the distance between positive sample pairs is too large and the distance between negative sample pairs is too small, allowing the network to learn an effective similarity measure.

The triplet loss function introduces three input samples: the anchor, the positive sample, and the negative sample. It further refines the relative relationship modeling between features by enforcing the constraint that the distance between the anchor and the positive sample is at least smaller than the distance between the anchor and the negative sample by a fixed margin. The specific formula is as follows:

$$L = \max(d_{ap} - d_{an} + m, 0) \tag{2.16}$$

where d_{ap} is the distance between the anchor and the positive sample, d_{an} is the distance between the anchor and the negative sample, and m is a predefined margin that controls the minimum difference in distance. This loss function ensures that the distance between the anchor and the positive sample is minimized, while the distance between the anchor and the negative sample is maximized, allowing for better discriminative feature learning.

The definition of positive and negative samples, along with the sampling strategy, plays a crucial role in the effectiveness of feature representation learning and the performance of the Siamese Network. For positive samples, various data augmentation techniques are commonly employed, such as rotation, scaling, cropping, lighting variation, and translation. These operations ensure that images of the same object, under different viewpoints, scales, and environmental conditions, are still considered positive sample pairs, thus enhancing the model's robustness to various external changes. For negative samples, the sampling strategy is more diverse. The most basic method is random negative sample sampling, where samples from different categories or instances are randomly selected as negative sample pairs. However, random sampling often leads to a large number of easy negative samples, which have significant differences from positive samples in the feature space. These samples contribute little to the training process, often resulting in gradient vanishing or premature convergence to suboptimal solutions.

To improve training efficiency and discriminative ability, hard negative mining strategies are typically employed. Specifically, hard negative samples are those that have a high similarity to the anchor sample in the feature space but belong to a different class. By selectively training on these "similar but different" negative samples, the learning difficulty of discriminative features is increased, prompting the network to learn more fine-grained and discriminative feature representations. Common implementations of this strategy include:

- **Offline Hard Negative Mining.** First, train a basic model in the initial phase, then calculate the feature representations of all samples in the entire dataset. The negative samples closest to the anchor features are selected to construct the training set.
- **Online Hard Negative Mining.** Dynamically select the most difficult negative samples (e.g., those with the closest feature distance to the anchor) within each training batch. This method is often used in conjunction with the triple loss.

In addition, the recently proposed batch-based in-batch negative sampling method has also been widely applied to accelerate training and enrich sample information. In this strategy, other samples within each training mini-batch (except for the positive pair) are automatically treated as negative samples. This enriches the negative sample information and accelerates model convergence without additional sampling overhead. This method has been shown to effectively increase the number of negative samples while reducing the cost of negative sample selection in contrastive learning methods, such as SimCLR [8] and MoCo [36]. To further enhance training effectiveness, some approaches have introduced temperature scaling to adjust the similarity distribution of negative samples or adopted a memory bank mechanism to extend the source of negative samples beyond the batch.

In visual object tracking, RGB-based, Thermal-based, LiDAR-based, and Event-based Siamese Networks generally follow a dual-branch "template–search" structure, but they differ significantly in feature extractor design, modality awareness, and temporal modeling. RGB-based Siamese Networks typically use shared-parameter CNNs to extract features from the template and search regions and perform matching via cross-correlation or Region Proposal Network (RPN [51]), excelling at capturing texture and color information with a standard structure. Thermal-based Siamese Networks are structurally similar to RGB-based ones, but due to the lack of color and fine details in thermal images, they often use shallower feature extractors and incorporate modality adaptation modules or non-shared branches for RGB-Thermal fusion tasks and cross-modality matching. LiDAR-based Siamese Networks are adapted for sparse point cloud input, employing geometry-aware networks that focus on structural rather than appearance matching, using spatial alignment and distance encoding to extract 3D features. Event-based Siamese Networks process sparse, asynchronous event streams by combining temporal convolution, event accumulation, or SNNs for feature extraction, and integrate LSTM or temporal attention to model dynamic patterns, with matching mechanisms designed to align in the temporal domain.

2.2.2 Typical Methods

In Siamese networks, Fully convolutional Siamese (SiamFC) [2] and Siamese Region Proposal Network (SiamRPN) [43] are two representative and typical methods. The following sections will use these two approaches as examples to systematically explain their network architectures and key algorithmic details.

SiamFC [2] proposed a fully convolutional Siamese network–based object tracking method, which innovatively formulates the target matching problem as a generic similarity metric learning task. The key strength of SiamFC [2] lies in its end-to-end offline training framework, which learns a generalizable similarity function between the target template and the search region using large-scale video datasets. This design eliminates the need for complex online procedures such as fine-tuning, incremental learning, or classifier retraining, enabling the model to perform target localization with a single forward pass per frame. As a result, the system achieves high speed, structural simplicity, and strong stability during tracking. Additionally, SiamFC [2] incorporates a multi-scale matching strategy by constructing a pyramid of search image scales to address target scale variation, further improving its adaptability and generalization. The introduction of this method laid the foundation for a series of Siamese-based trackers and has had a lasting impact on the field of VOT. As illustrated in the Fig. 2.5, the overall architecture consists of two input branches: the template image and the search image are cropped into patches of size

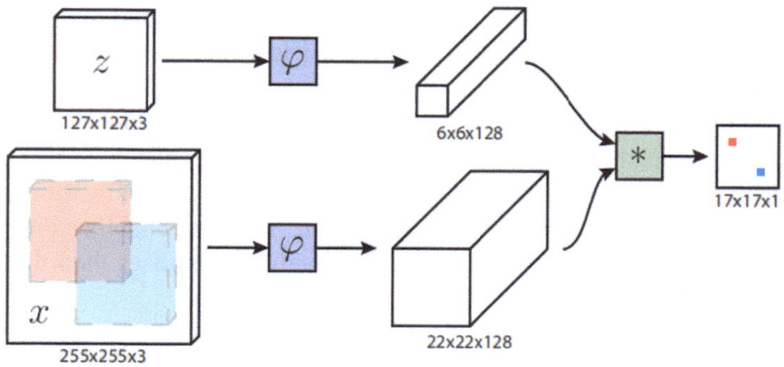

Fig. 2.5 Fully convolutional Siamese architecture [2]

127 × 127 and 255 × 255, respectively, representing the target appearance in the initial frame and the candidate region in the current frame.

In terms of network architecture design, SiamFC [2] adopts a symmetric Siamese structure in which both branches share weights and are based on a modified AlexNet [41] with fully connected layers removed, serving as the backbone feature extractor. To ensure that the model maintains a strictly fully convolutional nature, all padding operations are removed from the network, allowing the convolutional output to maintain consistent spatial receptive field progression, which facilitates dense spatial matching. The template image and the search image are passed through two completely symmetric convolutional paths to extract feature maps. These maps are then subjected to efficient cross-correlation operations, which involve sliding-window-level dot products between the two feature maps. This process generates a dense similarity score map over the search image, where each position indicates the degree of similarity to the target template. Finally, the position with the maximum value in the response map is regarded as the most probable location of the target in the current frame, thereby completing the target localization.

In terms of loss function design, SiamFC [2] adopts a logistic regression loss to supervise the similarity response map. Specifically, the central region of the response map is treated as positive samples, while the surrounding regions are considered negative samples. By maximizing the response values in the positive region and suppressing those in the negative region, the model achieves discriminative optimization of the similarity metric function. This dense supervision mechanism not only improves the matching accuracy but also accelerates the model convergence process.

However, despite its efficiency, the original SiamFC [2] framework exhibits limited discriminative capability, particularly in complex scenarios involving occlusion, distractors, or multiple similar objects. To address this limitation, SimFC [2] is proposed as a discriminative enhancement module that augments the standard Siamese pipeline. Its core idea is to introduce a lightweight feedforward classifier after the traditional correlation matching stage, allowing for more refined modeling of the similarity between candidate regions and the template. This significantly improves the model's ability to distinguish between foreground and background, especially in challenging scenarios such as occlusion, distractors, and multiple similar objects. SimFC [2] addresses this limitation by feeding the similarity vector at each position of the response map into a shared classifier for binary classification. This classifier, usually composed of one or more fully connected layers, outputs the probability of each location being the target center. Essentially, this transforms spatial similarity estimation into a classification problem, enabling the model to learn more complex decision boundaries. The method comprises the following key steps:

- **Feature Extraction**: A shared backbone network (e.g., deep convolutional neural network) is used to extract features from both the template and the search image.
- **Initial Similarity Computation**: A response map is generated by computing the cross-correlation or dot product between the template and search region features.
- **Similarity Fusion Classification**: For each location in the response map, its corresponding similarity vector is input into a fully connected classifier, which outputs the probability of the location being the target center.
- **Final Prediction**: The target location is inferred by selecting the location with the highest probability, followed by bounding box regression or scale adjustment if necessary.

Siamese Region Proposal Network (SiameseRPN [43]) is a high-performance, real-time VOT method. Its core contribution lies in integrating the traditional Siamese network structure with a Region Proposal Network (RPN) [51], enabling one-shot detection-based object tracking through end-to-end training. The overall network structure, as shown in the Fig. 2.6, consists of two functional modules: one is the Siamese feature extraction subnet with shared parameters, and the other is the Region Proposal Network (RPN) [51] used for target classification and position regression. This architecture overcomes the limitations of earlier Siamese methods, which only performed dense similarity matching, by incorporating precise localization of candidate bounding boxes into the end-to-end learning process. As a result, it significantly improves tracking accuracy and robustness while maintaining speed.

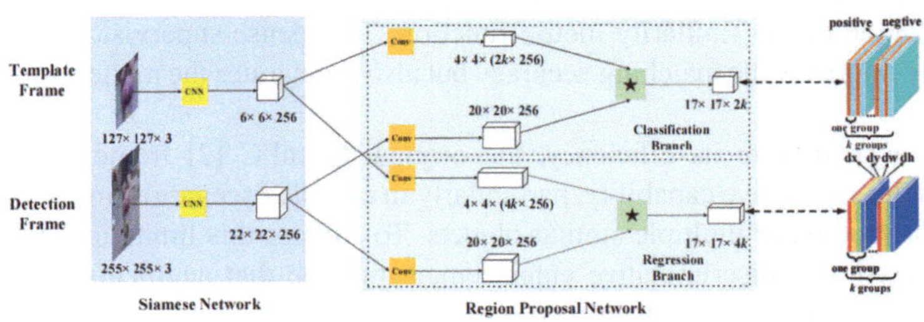

Fig. 2.6 Main framework of Siamese-RPN [43]

Specifically, the Siamese feature extraction subnet adopts a symmetric Siamese convolutional structure, receiving inputs from both the target template image (127 × 127) and the current frame search image (255 × 255). Both inputs share weights to maintain structural consistency and feature space alignment. The subnet is built on an improved version of the AlexNet [41] backbone, with grouped convolutions removed from the original architecture to enhance the network's cross-channel information interaction capability. It outputs high-level semantic feature maps of fixed dimensions. Building on this, the region proposal subnet fuses the feature maps of the template image and the search image through a cross-correlation operation, generating a joint feature map with semantic responses. The template image and the search image are processed through a twin-branch neural network with shared weights to extract their respective features. These features then interact in the feature space, typically via cross-correlation, attention mechanisms, or feature fusion modules, to model their similarity relationship. In this process, the template features serve as a query or filter applied to the search region features, generating a response map for target localization. This feature map is then passed through two parallel prediction branches: the classification branch and the regression branch. The classification branch outputs foreground/background classification scores for predefined anchor points and is supervised using a Softmax loss function. The regression branch predicts the spatial offsets between each anchor point and the ground truth bounding box, optimizing prediction accuracy with a smooth L1 loss. To improve the model's adaptability to target scale and deformation, SiameseRPN [43] employs a multi-scale anchor box strategy with single-scale anchors, thereby enhancing scale generalization while maintaining computational efficiency.

During the inference phase, SiameseRPN [43] models visual tracking as a local one-shot detection task. It extracts features from the target image only in the first frame to precompute convolutional kernels for the template branch and keeps the template features fixed throughout the tracking process. In subsequent frames, only the search region is processed, and cross-correlation is applied to generate target proposals efficiently. To improve computational efficiency and tracking stability, two proposal refinement strategies are adopted: (1) retaining only anchor points near the center of the classification feature map to reduce redundancy and (2) applying a cosine window and scale penalty to re-rank proposals, followed by non-maximum suppression (NMS) to select the final tracking box. This design enables the tracker to achieve real-time performance of up to 160 FPS with high accuracy and strong robustness, without relying on online updates or multi-scale testing. Furthermore, SiameseRPN introduces a temporally-aware RPN-like architecture that addresses the temporal inconsistency often overlooked in traditional Siamese frameworks. A temporal fusion module is integrated into the network to better capture temporal dynamics, while a candidate region scoring network jointly predicts classification confidence and localization accuracy. During training, a multitask objective combining logistic loss and regression loss is employed to maximize response at the true target location and suppress background noise, further enhancing the tracker's precision and robustness.

Moreover, SiameseRPN [43] combines Siamese networks and Region Proposal Networks (RPNs) to achieve end-to-end offline training and real-time tracking. The core framework consists of two parts: the Siamese feature extraction subnet and the region proposal subnet. The Siamese subnet contains two branches (template branch and detection branch), which process image patches from the history frame and the current frame, respectively, extracting features through a shared parameter convolutional network. The region proposal subnet uses correlation operations to generate classification and regression feature maps. The features of the template branch are used as convolutional kernels to perform cross-correlation with the features of the detection branch, generating object proposals. The classification branch outputs activation maps for foreground and background, while the regression branch outputs the offsets for bounding boxes. End-to-end training is achieved through joint optimization of the classification loss (cross-entropy) and regression loss (smooth L1).

- The core of the Siamese-RPN [43] framework is the use of correlation between the template and detection feature maps to produce classification and regression outputs. This is implemented as group-wise convolution:

$$\begin{aligned} A^{cls}_{w\times h\times 2k} &= [\phi(x)]^{cls} \star [\phi(z)]^{cls} \\ A^{reg}_{w\times h\times 4k} &= [\phi(x)]^{reg} \star [\phi(z)]^{reg} \end{aligned} \tag{2.17}$$

  where $\phi(z)$ and $\phi(x)$ are the template and detection feature maps, respectively, and $\star$ denotes the cross-correlation operation.
- To regress the object location and size, the network minimizes the distance between anchors and ground-truth boxes using the following parameterization:

$$\begin{aligned} \delta[0] &= \frac{T_x - A_x}{A_w}, \quad \delta[1] = \frac{T_y - A_y}{A_h}, \\ \delta[2] &= \ln\left(\frac{T_w}{A_w}\right), \quad \delta[3] = \ln\left(\frac{T_h}{A_h}\right) \end{aligned} \tag{2.18}$$

  where (A_x, A_y, A_w, A_h) and (T_x, T_y, T_w, T_h) represent the center coordinates and dimensions of the anchor and the ground truth box, respectively.

In summary, SiameseRPN [43] maintains the simplicity and parallelism of the Siamese architecture while incorporating a region proposal mechanism and explicit bounding box regression strategy. This significantly extends the modeling capabilities of Siamese networks for object-tracking tasks and lays the foundation for the development of subsequent detection-based tracking methods. Its end-to-end modeling approach and optimization strategies for inference efficiency are of significant reference value in both real-time industrial applications and academic research.

In the VOT task, Siamese networks have become a mainstream framework due to their balanced accuracy and speed, leading to the emergence of many variants. In addition to the previously mentioned SiamFC [2] and SiamRPN [43], there is CFNet [54], which integrates correlation filtering as a differentiable layer into deep neural networks, achieving state-of-the-art tracking performance in high frame rates with a lightweight architecture. SiamRPN++ [13] uses discriminative training to learn similarity metrics, employing a Siamese convolutional network architecture to map inputs to a low-dimensional space, ensuring that similar samples are close and dissimilar samples are far apart. DaSiamRPN [69] significantly improves tracking accuracy and robustness by improving the training data distribution and introducing an interference-aware module, achieving real-time high-performance in both short-term and long-term tracking tasks. SiamMask [57] combines object tracking and video object segmentation into one framework through multitask learning, generating real-time segmentation masks and rotated bounding boxes for the target. SiamBAN [11] directly predicts the target bounding box through an adaptive region proposal network without requiring predefined anchor boxes. TransT [10] is a vision-based object tracking method based on the Transformer attention mechanism, which directly integrates template and search region features using self-attention (ECA) and cross-attention (CFA) modules. SA-Siam [33] is a dual Siamese network consisting of a semantic branch and an appearance branch, where the semantic branch uses deep features from a pretrained classification network and introduces a channel attention mechanism, and the appearance branch employs a Siamese network structure for similarity learning. After independent training of both branches, real-time object tracking is achieved through weighted response map fusion.

CGACD [26] proposed a two-stage corner point detection and tracking method that integrates pixel-level and channel-level correlation-guided attention mechanisms. By fully exploiting the correlation between the template and candidate regions, this approach significantly enhances the accuracy of corner point detection and bounding box estimation in object tracking, while maintaining high-speed performance. SiamGAT [31] constructs a complete bipartite graph between the template and the search region and introduces a graph attention mechanism to achieve part-based feature association and target-aware region selection. SiamCAR [32] proposes a fully convolutional Siamese classification and regression tracking framework that is anchor-free and proposal-free. It reformulates the visual tracking task as per-pixel category prediction and bounding box regression, enabling end-to-end joint learning of similarity features and spatial information. SiamRN [12] introduces a meta-learning-based Relation Detector (RD) module to learn how to suppress background distractions in the feature space. This is coupled with a Refinement Module (RM) for joint optimization of the classification and regression branches, significantly improving the discriminative power and robustness of the tracker in complex scenarios.

2.2.3 Summary and Challenges

The parameter-sharing mechanism in Siamese Networks enhances computational efficiency: the twin-branch architecture with shared weights significantly reduces the number of model parameters and mitigates the risk of overfitting commonly associated with independently parameterized dual-branch networks. Furthermore, the Pseudo-Siamese Network [60] modifies the architecture to accommodate multi-modal inputs, demonstrating strong performance in cross-modal retrieval and image-text matching scenarios.

Although Siamese networks have demonstrated impressive performance in VOT tasks, with advantages such as a simple architecture and efficient matching, they still face several limitations in complex real-world scenarios. Future research can focus on multiple directions to enhance their capabilities. Firstly, current approaches are limited in modeling high-level semantic and contextual dependencies, making them less effective in scenes involving drastic appearance changes or ambiguous targets. The incorporation of Transformer architectures or multi-scale attention mechanisms could significantly improve feature representation and discriminative power. Secondly, to enhance robustness against occlusions, illumination changes, and background interference, strategies such as dynamic template updating, adaptive weight fusion, or the integration of long-term memory modules may offer more stable target representations.

From a deployment perspective, balancing accuracy with inference speed remains a critical research direction, especially for real-time applications requiring low latency and high frame rates. Techniques such as lightweight network design, knowledge distillation, and model pruning can help reduce computational costs without sacrificing performance. Additionally, multi-modal information fusion—integrating RGB, infrared, and depth modalities—has the potential to deliver more robust tracking under extreme conditions. End-to-end joint optimization frameworks, along with unsupervised or weakly supervised learning approaches, could further improve adaptability while reducing reliance on annotated data.

2.3 Discriminative Correlation Filters

2.3.1 Paradigm Introduction

Discriminative Correlation Filters (DCFs) are an efficient object-tracking method that trains a filter in the frequency domain to distinguish between the target and the background, thereby enabling fast and accurate tracking. The DCF approach leverages the circulant structure of training data to augment the number of samples synthetically and employs convolution in the frequency domain to significantly accelerate computation. This design allows DCF-based trackers to maintain high accuracy while achieving real-time performance. The core idea is to learn a filter on a large set of synthesized samples such that the filter produces a strong response at the target location and weak responses elsewhere, enabling precise target localization. The fundamental framework of DCF can be summarized in the following key steps:

- **Training Sample Construction:** Given a target image patch from the current frame, the model generates a large number of training samples through a cyclic assumption or circular shifting. This method enriches the training data while maintaining computational efficiency. Without requiring explicit sampling, it theoretically provides dense coverage of all possible spatial translations, significantly expanding the effective size of the training set.
- **Target Regression Design:** In the target regression design, each training sample is assigned a regression label in the form of a Gaussian response map. The filter is trained to produce the highest response at the target location. Typically, the center of the template corresponds to the peak response, while the reaction gradually decays with increasing distance from the center, and background regions receive minimal response values. This enhances the model's localization capability. The Gaussian response can be defined as:

$$y(\mathbf{x}) = \exp\left(-\frac{\|\mathbf{x} - \mathbf{x}_0\|^2}{2\sigma^2}\right) \tag{2.19}$$

  where σ controls the spatial spread of the response.
- **Filter Learning:** In this step, the filter weights are learned by minimizing the mean squared error (MSE) between the predicted response and the Gaussian label, with an additional regularization term to prevent overfitting.
- **Detection and Inference:** During the inference stage, features are extracted from a candidate region in a new frame and correlated with the trained filter to produce a response map. The position of the maximum value in the response map is identified as the new target center, and the target bounding box is updated accordingly.
- **Online Update:** To adapt to the target's appearance changes over time, the filter is incrementally updated with each frame. A common update strategy is the Exponential Moving Average (EMA), defined as:

$$h_{\text{updated}} = (1 - \eta) \cdot h_{\text{old}} + \eta \cdot h_{\text{new}} \tag{2.20}$$

  where η is the learning rate, h_{new} is the filter learned from the current frame, and h_{old} is the filter from the previous frame.

Discriminative Correlation Filters have garnered significant attention in computer vision and pattern recognition research. Their architectural design incorporates several key features to enhance the model's representational capacity and generalization performance. The following discusses these aspects from the perspectives of end-to-end training, multi-scale processing, and online learning mechanisms:

- **End-to-End Training.** Discriminative Correlation Filters (DCFs) adopt an end-to-end training paradigm, where the correlation filters and discriminative functions for the target are learned directly from raw input data through a unified optimization process. In this framework, the filter learning module and the target discrimination module are no longer modeled independently; instead, they are functionally coupled and jointly optimized, enabling adaptive learning tailored to the needs of visual tracking tasks. Compared to traditional approaches that rely on handcrafted or statically pre-trained features, this strategy significantly reduces information loss and the semantic gap introduced by feature engineering, thereby enhancing the overall representational capacity and tracking performance.
  Moreover, the end-to-end architecture provides a unified and efficient modeling platform for multitask tracking. Discrimination tasks across multiple targets or scales can be jointly trained on shared low-level feature representations, which not only reduces computational redundancy but also improves robustness and adaptability through implicit regularization across tasks.
- **Multi-Scale Processing.** In visual tracking tasks in computer vision, the target object often exhibits significant scale diversity, including both large-scale motions of the main target and small targets with blurred boundaries and rich details. This multi-scale characteristic presents a higher challenge to the Discriminative Correlation Filter (DCF) model's discriminative capability across different spatial scales. To address this, DCF methods have gradually incorporated multi-scale processing mechanisms to improve their ability to perceive and match targets at various scales. Currently, mainstream improvement strategies often employ explicit multi-scale feature fusion methods. Typical examples include Feature Pyramid Networks (FPN [46]) and High-Resolution Networks (HRNet [56]), which are integrated into the DCF framework to enhance its response capability across different scales. These structures establish lateral connections and hierarchical fusion operations between multi-level features, enabling collaborative modeling of local details and global contextual information. In addition, recent studies have proposed dynamic scale sampling mechanisms, which, when combined with DCF models, adaptively adjust the processing scale based on the target features in an image frame. This approach enhances the flexibility and efficiency of scale estimation. To further improve the DCF model's adaptability to complex spatial deformations and scale inconsistencies, some research has introduced mechanisms like Deformable Convolutions [16] to dynamically adjust the receptive field of correlation filters.
- **Online Learning Mechanism.** To address challenges in practical target tracking applications, such as data distribution shifts, target appearance drift, and imbalanced class distributions, modern Discriminative Correlation Filter (DCF) models have gradually introduced online learning mechanisms, enabling them to continuously and adaptively update. This mechanism allows the filter to dynamically receive newly emerging target appearance information without relying on complete historical frame data and continually optimize the correlation filter parameters in the process. As a result, the model's adaptability and practicality

in non-static scenarios are significantly enhanced. Building on this, Incremental Class Learning (ICL) has gradually been integrated into the DCF framework as a key online learning task in target tracking. ICL requires the DCF model to maintain its discriminative ability for previously learned target features when new targets or changes in target states appear, thus avoiding the issue of catastrophic forgetting. To achieve this goal, current research tends to combine strategies such as discriminative boundary regularization and knowledge distillation within the DCF model, ensuring explicit retention and constraint of previous target representations while online updating the filter.

From a methodological perspective, Discriminative Correlation Filters (DCFs) exhibit a range of structural and computational advantages, making them a foundational paradigm in real-time VOT since their inception. Over time, this framework has inspired numerous variants and extensions. The core design principle of DCF lies in its mathematical restructuring, which enables efficient and real-time discriminative model learning and prediction. The key advantages are as follows: First, DCF transforms the training samples into a circulant matrix structure by introducing a periodic assumption at the modeling level. This allows the convolution operation in the spatial domain to be converted into element-wise multiplication in the frequency domain, based on the convolution theorem from Fourier analysis. This transformation greatly simplifies the filter optimization process. Specifically, it enables the use of the Fast Fourier Transform (FFT) for efficient forward propagation and gradient updates, resulting in a closed-form analytical solution and avoiding complex numerical optimization procedures. This mechanism significantly reduces computational complexity, allowing the model to perform training and inference within milliseconds, thereby meeting the stringent real-time requirements of visual tracking tasks.

Secondly, DCF methods typically adopt a ridge regression framework for model training, where an L2-norm regularization term is introduced into the objective function. This effectively suppresses overfitting and enhances the numerical stability and generalization capability of the model. The framework offers a clear analytical structure, which facilitates theoretical analysis and performance evaluation, while also providing a unified optimization basis for subsequent extended models. At the same time, the structural design of DCF models is highly concise, usually employing a linear filter form. This avoids excessive network stacking and high-dimensional nonlinear mappings, offering strong interpretability and ease of engineering implementation. Moreover, DCF methods inherently support online learning. Specifically, filter weights can be updated frame-by-frame using strategies such as sliding averages or exponential weighting, allowing the model to dynamically adapt to changes in the target's appearance during tracking. This characteristic grants DCF a certain degree of robustness under non-ideal conditions such as target scale variation, occlusion, and illumination changes. It also lays a solid foundation for further research that incorporates appearance modeling, attention mechanisms, and temporal structures.

In visual object tracking (VOT) tasks, Discriminative Correlation Filter (DCF) methods are widely adopted in single-object, single-initialization frameworks due to their superior computational efficiency and structural simplicity. The core idea is to construct a discriminative filter based on the initial target annotation in the first frame, enabling the model to distinguish the target from the background. This filter is continuously updated in subsequent frames to adapt to changes in the target's appearance. The entire technical pipeline can be divided into two main stages: the training stage and the inference stage, which together constitute the complete operational mechanism of DCF in tracking scenarios.

- **Filter Training Stage.** In the initial frame, once the target's bounding box annotation is provided, the Discriminative Correlation Filter (DCF) method crops the target image patch as a template region. Through circular shifting operations in the spatial domain, numerous synthetic training samples are constructed, leveraging the periodicity assumption to encode all possible translations of the image patch into a circulant matrix. This formulation significantly reduces data dimensionality and computational complexity by transforming the enumeration of multiple samples into matrix operations.
  DCF then formulates a least-squares regression model to learn the filter f, generating an ideal response map y centered on a Gaussian distribution, where the response peaks at the target location and decays gradually elsewhere. The final objective becomes a regularized least squares problem (i.e., ridge regression), expressed as:

$$\min_f \|Xf - y\|^2 + \lambda \|f\|^2 \tag{2.21}$$

  Here, x denotes the input template feature, and $*$ represents the circular convolution operation. The term λ is the regularization coefficient used to control model complexity and prevent overfitting.
  To accelerate computation, DCF methods exploit the equivalence of circular convolution in the frequency domain. Specifically, convolution operations are transformed into element-wise multiplication through the Fast Fourier Transform (FFT), allowing the optimal filter to be efficiently obtained in the frequency domain. Based on the convolution theorem of Fourier analysis, the solution can be derived in a closed-form expression as follows:

$$\hat{f} = \frac{\hat{x} \odot \hat{y}^*}{\hat{x} \odot \hat{x}^* + \lambda} \tag{2.22}$$

  where $\hat{f}$, $\hat{x}$, and $\hat{y}$ are the Fourier transforms of the filter, input feature, and target response respectively; $\odot$ denotes the element-wise (Hadamard) product, and * represents the complex conjugate. The regularization term λ helps prevent overfitting.

- **Inference and Localization.** During each frame of the tracking process, DCF utilizes the learned filter f to perform correlation over the current search region image z, formulated as:

$$r = f * z \tag{2.23}$$

where r denotes the response map, representing the similarity between the target template and candidate positions in the search region. The position corresponding to the maximum value in the response map is typically taken as the new target location, enabling inter-frame localization and tracking.
To improve adaptability to scale variations, many DCF variants adopt a multi-scale search mechanism. This involves repeating the correlation process over multiple resized versions of the search region and selecting the scale corresponding to the highest response value as the current target size.
- **Online Update Mechanism.** Considering that the target appearance may change over time due to factors such as illumination, pose variation, and occlusion, DCF methods incorporate an online model update mechanism during inference. This mechanism constructs new training terms from each frame by extracting the image patch features x_t and the corresponding ideal response y_t and updates the filter parameters using a sliding average or exponential weighting scheme:

$$f_t = (1 - \eta) f_{t-1} + \eta f_{\text{new}} \tag{2.24}$$

where $\eta \in [0, 1]$ is the learning rate that controls the balance between the old and new models. This updated strategy preserves temporal continuity and enhances robustness, enabling the filter to adapt dynamically to gradual changes in target appearance and improving tracking stability.

However, the standard DCF method introduces unrealistic background information at the boundaries due to its periodic assumption, which affects the tracking accuracy. To address this, subsequent works introduced a series of improvements, such as incorporating spatial weights (e.g., CSR-DCF [68] or cropping operations (e.g., BACF [40]) to alleviate the boundary effect. Additionally, extended versions such as KCF [38] (Kernelized Correlation Filters) extend DCF to the kernel method framework, performing filter learning in a nonlinear feature space, which further enhances tracking discriminability. In practical applications, DCF-based methods have also introduced multi-channel features to improve the model's adaptability to appearance changes and employ online updating strategies to perform frame-by-frame adaptive adjustments of the filter, dynamically adapting to changes in target scale, pose, and appearance.

In visual tracking, Discriminative Correlation Filters (DCFs) share a common framework of "filter learning + response map generation" across RGB, thermal infrared, LiDAR, and event-based modalities, but their network structures differ significantly due to modality-specific characteristics. RGB-based DCF uses pre-trained CNNs to extract deep appearance features, combined with frequency-domain convolution or end-to-end filter learning, making it suitable for texture-rich scenes; Thermal-based DCF employs shallower feature extractors and incorporates modality-adaptive modules (such as attention mechanisms or normalization layers), in RGB-Thermal fusion, a dual-branch structure is often used to merge features and enhance discriminative capability under thermal imaging; LiDAR-based DCF first converts sparse point clouds into BEV or depth maps, then applies geometry-aware networks to extract structural features, and uses spatially-aware filters to ensure consistency in 3D tracking; Event-based DCF accumulates event streams into pseudo-frames or voxel grids, uses CNNs combined with temporal modeling to extract dynamic features, and builds time-aware filters to handle fast motion and motion blur effectively. Each modality thus adapts the DCF structure in feature extraction, modality alignment, and temporal processing to best suit its perceptual environment.

2.3.2 *Typical Methods*

Among correlation filter-based trackers, Accurate Tracking by Overlap Maximization (ATOM) and Discriminative Model Prediction (DiMP) are two representative classical methods. In the following, we systematically elaborate on their network architectures and core algorithmic mechanisms using these two methods as examples.

Accurate Tracking by Overlap Maximization (ATOM) [20] is a high-precision tracking framework proposed for general VOT tasks. The network structure diagram is shown in Fig. 2.7. Its significant contribution lies in the explicit decoupling of object classification and state estimation tasks at the architectural level. This design aims to overcome the precision bottleneck caused by directly using classification

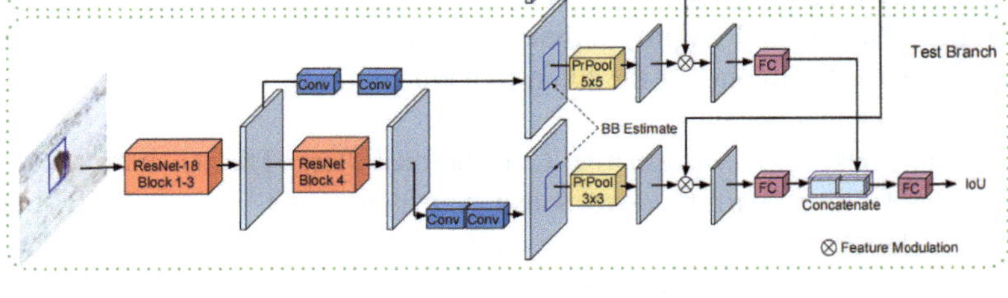

Fig. 2.7 ATOM Network Architecture Diagram [20]

responses for bounding box localization in traditional Discriminative Correlation Filter (DCF) methods, especially when dealing with issues such as target scale variation, deformation, rotation, and performance degradation caused by complex background interference.

ATOM [20] consists of two collaboratively functioning branches: the online object classification branch and the offline trained bounding box estimation branch. The online classification branch is responsible for discriminating the target and background information in the current image region, providing a coarse-grained target location estimate. The state estimation branch, based on this classification result, further refines the prediction to achieve high-precision target bounding box estimation. In terms of object classification, ATOM [20] constructs a compact, fully convolutional neural network classifier, which continuously adjusts its weight parameters through online optimization in each frame to maintain the persistent discriminative capability for the target. The optimization objective is a Gaussian-form regression loss function, aiming to maximize the response value at the target center while suppressing responses in non-target regions. To achieve efficient online updates, ATOM [20] adopts a combined optimization strategy that integrates the Gauss–Newton method with the conjugate gradient method, significantly reducing computational load while maintaining good convergence properties and adapting to the high-frequency online update requirements of deep neural network structures. This method strikes a balance between efficiency and flexibility, offering good convergence and framework adaptability, and can be directly implemented in deep learning platforms without the need for manual derivation of complex gradients. ATOM [20] cleverly combines the high-level semantic modeling capability of offline learning with the real-time adaptability of online learning, where the offline-trained target estimation module captures general target appearance variations using large-scale datasets, while the online-trained classifier dynamically adapts to background interference and target appearance changes in real-world tracking scenarios.

In terms of state estimation, ATOM [20] introduces a regression branch based on an IoU prediction mechanism, which fundamentally improves upon traditional methods that rely on maximum response point localization strategies. This module uses a regression network pre-trained offline on large-scale video datasets. The input consists of candidate bounding boxes and the feature map of the current frame, while the output is the predicted IoU value between each candidate box and the true target. To enhance the target specificity of the predictions, the branch employs a feature modulation mechanism, injecting target appearance information extracted from the initial reference frame into the candidate box scoring process. Finally, a differentiable gradient ascent optimization strategy is used to fine-tune the candidate bounding box parameters in the continuous space, to locate the position with the maximum IoU, and thereby produce an accurate target state estimate. In contrast, traditional DCF series methods often combine the classification and state estimation processes into a single model, directly determining the target position based on the location of the maximum value in the response map. This approach struggles to model the geometric shape variations of the bounding box. Therefore,

ATOM [20] achieves a key breakthrough in its design by separating tasks and reconstructing the optimization objectives, leading to more expressive and stable tracking performance.

On the basis of the above core concepts, ATOM [20] further decomposes the tracking task into two specialized components: target estimation and target classification. The estimation component, discussed above, achieves high-precision bounding box regression through an offline-trained IoU prediction network. The object estimation component achieves high-precision bounding box regression via an offline-trained IoU prediction network, while the object classification component differentiates the target from background interference through an online-optimized, lightweight discriminative model. In terms of object estimation, ATOM [20] employs a modulated IoU prediction network architecture. The network consists of two branches: the reference branch processes the reference frame's features and target box, outputting a modulation vector, and the test branch processes the current frame's features and candidate boxes, extracting features via Precise ROI Pooling (PrPool). The outputs of these two branches are modulated and fed into a three-layer fully connected network to predict the IoU overlap between the current candidate box and the true target.

The target classification module is a lightweight two-layer fully convolutional network designed to distinguish between the target and background interference. The target classification component adopts a simple two-layer fully convolutional network structure. The first layer uses a 1×1 convolution for feature dimensionality reduction, while the second layer uses a 4×4 convolution to output a single-channel confidence map. To achieve fast online optimization, ATOM [20] innovatively proposes a conjugate gradient optimization algorithm based on Gaussian–Newton approximation, replacing traditional gradient descent. The residual is defined as follows:

$$r_j(w) = \sqrt{\gamma_j}(f(x_j; w) - y_j), \quad r_{m+k}(w) = \sqrt{\lambda_k} w_k \tag{2.25}$$

$$\ell_w(\Delta w) = \frac{1}{2}\| J_w \Delta w + r(w)\|^2 \tag{2.26}$$

where J_w is the residual Jacobian matrix at parameters w.

$$\mathcal{L}(w) = \sum_{j=1}^{m} \gamma_j \| f(x_j; w) - y_j\|^2 + \sum_{k=1}^{n} \lambda_k w_k^2 \tag{2.27}$$

- The algorithm leverages modern deep learning frameworks to automatically compute Jacobian matrices via backpropagation, eliminating the need for error-prone manual derivation.
- The training process uses a weighted L^2 loss function (as shown above) with regularization terms λ_k.

In summary, the ATOM [20] method, while retaining the advantage of online updates in discriminative trackers, introduces an offline state estimation module with strong expressive power. Through unified modeling of the optimization strategy and end-to-end implementation of the computational path, ATOM achieves excellent performance on multiple standard visual tracking datasets. Notably, it shows significant advantages over concurrent methods, especially in terms of bounding box estimation accuracy, adaptability to target deformations, and robustness to occlusions.

Traditional Discriminative Correlation Filter (DCF) methods have made significant progress in the field of visual tracking. The core idea is to learn a discriminative filter by minimizing the difference between the predicted response and the ideal Gaussian response, based on the target appearance from the initial frame during the testing phase. However, traditional methods have two fundamental limitations:

1. The optimization process is static, making it difficult to adapt flexibly to changes in the target appearance.
2. The training process typically relies on closed-form frequency domain solutions, making it challenging to integrate into end-to-end deep learning frameworks.

Discriminative Model Prediction (DiMP) [3] represents a fundamental improvement to the traditional DCF framework, particularly in terms of the target model learning mechanism, with its network architecture illustrated in the Fig. 2.8. This method constructs a prediction architecture that retains the advantages of discriminative learning while enabling dynamic prediction and end-to-end training of the target model. Specifically, DiMP [3] transforms the target classification model learning process into an explicitly modeled optimization process: the input is a sample set, and the output is the parameters of a convolutional classifier (i.e., the filter) used for distinguishing between the target and background regions. Unlike traditional DCF, DiMP [3] is no longer limited to a fixed least squares loss function but introduces a learnable discriminative loss structure, which includes target mask functions, spatial weighting terms, and label regression functions, all of which are automatically learned from data by the neural network, greatly improving the model's adaptability and discriminative capability. In terms of optimization strategy,

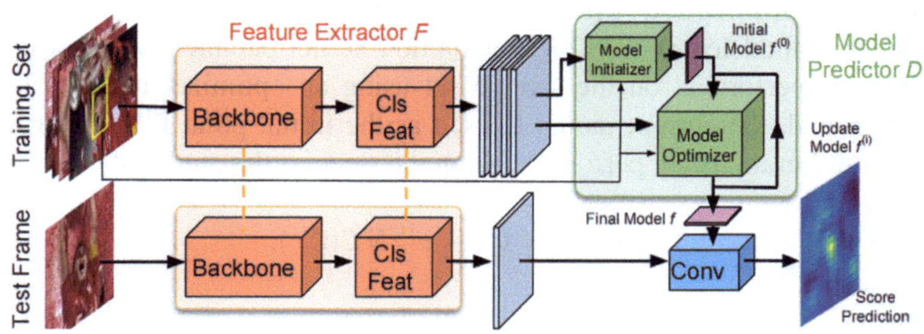

Fig. 2.8 DiMP Network Architecture Diagram [3]

DiMP [3] designs a filter update mechanism based on Steepest Descent, utilizing the current residual gradient and its Jacobian matrix to dynamically calculate the step size, significantly improving the optimization convergence speed. To further enhance inference efficiency, the method also introduces a lightweight initial model prediction module to estimate reasonable initial values before optimization begins.

Additionally, DiMP [3] introduces a refined online update mechanism. During the tracking process, the algorithm continuously maintains a finite-size online training sample set, which consists of high-confidence prediction results from previous frames. When update conditions are met, the current model is used as an initial point, and several steepest descent optimization iterations are performed on the new sample set to efficiently update the model. This mechanism not only allows for continuous adaptation to target appearance changes but also effectively avoids the information loss problem introduced by the traditional template averaging strategy. In VOT tasks, since the target's appearance is usually only labeled in the first frame, how to efficiently adapt to appearance changes and background disturbances in subsequent frames becomes a key factor influencing tracking performance. Although traditional discriminative correlation filter methods have some online learning capability, they are limited by fixed analytical solutions and the linear model's expressive power, making it difficult to fully adapt to complex dynamic scenes. To address this, DiMP [3] introduces an explicitly modeled optimization-based online learning mechanism. By integrating the optimization process into the network architecture, the model can dynamically predict and update the target classifier during the inference stage, significantly enhancing the model's adaptability and discriminative ability.

1. **Online Maintenance of the Sample Set.** In DiMP [3], during the tracking process, a training sample set $\mathbf{S}_{\text{train}}$ is continuously maintained for updating the target model. The set consists of feature maps $\mathbf{x}_j$ obtained from previous frames and their corresponding target center positions $\mathbf{c}_j$. The selection of samples is based on the confidence of the target detection in the current frame. If the detection response is significant, the sample from that frame is added to the set. To control computational cost, the sample set's capacity is limited to a fixed upper bound, and any excess samples are removed using a first-in, first-out (FIFO) strategy.
2. **Iterative Optimization Update of the Discriminative Model.** The target classifier is modeled as a set of convolution kernel parameters $\mathbf{f}$, and iterative learning is performed on the sample set using an optimizer module. In each update round, the current model is used as the initial point $\mathbf{f}^{(0)}$, and a limited number of iterations are performed using the steepest descent method:

$$\mathbf{f}^{(i+1)} = \mathbf{f}^{(i)} - \alpha^{(i)} \nabla \mathcal{L}(\mathbf{f}^{(i)}), \tag{2.28}$$

where $\alpha^{(i)}$ is the adaptive step size at the current step, and the loss function's second-order approximation at the current estimate point is derived (e.g., Gaussian–Newton method). This optimization process takes full advantage of the structural differences between positive and negative samples. Additionally, it enables the network to learn the essential components of the discriminative loss during the offline training phase. These components include the regression label function y_c, the target region mask m_c, and the spatial weighting function v_c, enabling finer-grained differentiation between target and background regions.

3. **Multi-strategy Update Scheduling Mechanism.** To achieve an effective balance between performance and computational efficiency during tracking, DiMP [3] designs a hierarchical model update trigger mechanism. The periodic update strategy triggers a complete target model optimization process at fixed frame intervals to continuously adapt to the gradual changes in the target's appearance. In addition, the system introduces a response-driven dynamic update mechanism: when multiple high-response regions appear in the response map, indicating potential interference peaks, a lightweight iterative optimization is immediately executed to improve the model's ability to discriminate interference targets and suppress drift risks. Furthermore, during the tracking initialization phase, after the target label is obtained in the first frame, the system will perform multiple rounds of iterative optimization to fully exploit the initial target appearance features, constructing a discriminative initial target model that lays a solid foundation for stable tracking in subsequent frames.

In terms of bounding box regression, DiMP [3] adopts the IoU maximization framework from ATOM [20], performing gradient optimization on the predicted box coordinates to maximize the predicted IoU value. This enables high-precision target localization and bounding box estimation. DiMP [3] has achieved a paradigm shift from static filter learning to an optimizable model prediction, inheriting the discriminative advantages of DCF methods, while significantly enhancing tracking adaptability and robustness through the design of a learnable optimization process and online update mechanism.

The core idea of the end-to-end trainable VOT framework proposed by DiMP [3] is to combine discriminative learning loss with efficient optimization strategies. This approach allows the model to produce a highly discriminative target representation within just a few iterations. Compared to traditional Siamese trackers, DiMP [3] can fully exploit the appearance information of both the target and background and achieve generalization to unseen targets by learning key parameters of the loss function itself. Specifically, the DiMP [3] algorithm uses a composite residual function composed of hybrid regression and hinge loss, expressed as:

$$r(s, c) = v_c \cdot \big(m_c s + (1 - m_c)\max(0, s) - y_c\big) \tag{2.29}$$

Among them, the target mask m_c, spatial weight v_c, and regression target y_c are parameterized by the Radial Basis Function (RBF) and are learned using a data-driven approach to achieve more accurate error modeling and more robust discriminative ability.

In the model prediction process, the algorithm employs an optimization strategy based on the steepest descent, efficiently estimating the optimal step size through the Gauss–Newton approximation. The calculation formula is as follows:

$$\alpha = \frac{\nabla L(f^{(i)})^\top \nabla L(f^{(i)})}{\nabla L(f^{(i)})^\top Q^{(i)} \nabla L(f^{(i)})} \tag{2.30}$$

Here, L denotes the loss function, $f(i)$ is the model prediction at the i-th iteration, and $Q(i)$ is the corresponding second-order approximation matrix. Thanks to this optimization strategy, the model typically achieves rapid convergence within 5–10 iterations, significantly enhancing the real-time performance and accuracy of online tracking.

The overall architecture of DiMP [3] mainly consists of two key components: the ROI pooling-based model initialization module, which extracts and estimates the initial target features to provide excellent initialization conditions; and the differentiable optimization module, which allows for joint training of the feature extractor and loss function parameters through the standard backpropagation algorithm, enabling end-to-end collaborative learning from feature representation to the optimization process. During the actual online tracking phase, DiMP [3] introduces a dynamic memory bank mechanism, maintaining a training set that includes the most recent 50 samples. Through periodic optimization iterations, it continuously updates and adapts to changes in the current target, thereby enhancing the robustness and adaptability of tracking.

MOSSE [4] proposed an adaptive correlation filter based on the minimum output squared error and MOSSE, optimizing the filter in the Fourier domain to minimize the error between the actual output and the Gaussian expected response, achieving efficient and robust VOT. CSK [37] introduced an efficient tracking method based on a circulant matrix structure and Fourier transform, achieving fast training and detection through dense sampling strategies and kernel techniques. By leveraging the properties of circulant matrices, the kernel correlation filter is extended to nonlinear spaces, achieving a closed-form solution in the Fourier domain. SAMF [44] proposed a scale-adaptive object tracking algorithm based on the kernel correlation filter framework. Through a multi-scale search strategy and feature fusion, it effectively addresses the limitations of traditional correlation filters with fixed template sizes. The algorithm uses the circulant matrix property to efficiently compute responses in the Fourier domain and implements scale estimation via bilinear interpolation. DSST [22] introduced a scale-adaptive tracking method based on discriminative correlation filters, which learns independent correlation filters for translation and scale estimation, explicitly modeling the appearance changes caused by target scale variations. STC [66] models the statistical correlation between the target and the low-level features of the local context region using a Bayesian framework and maximizes the likelihood function of the target's position through a confidence map. The algorithm employs a fast Fourier transform for efficient convolution operations and combines the time update mechanism of the spatial context model to effectively handle challenges such as occlusion, lighting changes, and scale variations.

CN [19] proposed a real-time visual tracking method based on color attributes by combining color name features with kernelized correlation filters (CSK [37]) and introducing adaptive dimensionality reduction techniques. This significantly improves tracking accuracy and speed. LCT [48] uses temporal context correlation to construct a regression model for robust translation estimation while training a discriminative correlation filter for scale estimation based on high-confidence frames. Additionally, a random fern classifier is used for target re-detection to handle tracking failures. The algorithm adopts a dual-regression model strategy with a fast-updating context regressor and a conservatively updating target regressor, combined with an online detection mechanism to effectively address the stability–plasticity dilemma in model updates. C-COT [21] uses an implicit interpolation model to fuse multi-resolution feature maps and employs Fourier domain optimization for sub-pixel accurate localization. This method generates a target confidence map through continuous convolution operations and solves a regularized least-squares problem using the conjugate gradient method. It achieves performance improvements in both object tracking and feature point tracking tasks. ECO [23] introduces factorized convolution operators to significantly reduce the number of model parameters and designs a compact generative sample space model to lower computational complexity. It also employs a sparse model update strategy to enhance robustness. The algorithm jointly optimizes convolution kernel decomposition, sample distribution modeling, and update mechanisms, effectively addressing the core issues of overfitting and low computational efficiency in traditional DCF methods. LADCF [63] alleviates boundary effects and filter degradation problems through adaptive spatial feature selection and temporal consistency constraints. The algorithm uses structured spatial sparsity constraints for feature selection and combines the Augmented Lagrangian Method (ADMM) for joint optimization, achieving spatiotemporal consistent filter learning in a low-dimensional discriminative manifold.

2.3.3 Summary and Challenges

DCF enables end-to-end optimization with a large number of negative samples (i.e., background regions), strengthening the filter's discriminative ability and allowing the tracker to better distinguish the target from distractors in complex environments. The DCF framework is also highly flexible and extensible—subsequent research has integrated multi-channel features (e.g., HOG [17]), spatial regularization, scale estimation, and appearance update mechanisms to further improve tracking performance. Nevertheless, DCF still faces several challenges. The circulant assumption introduces boundary effects, which can result in incorrect filter responses near the image borders, thereby affecting tracking accuracy. During the model update process, DCF can be overly sensitive to appearance changes, which may lead to model drift, especially in cases of occlusion, severe background interference, or temporary disappearance of the target. Variants such as SRDCF [24] and C-COT

[21] have addressed these limitations by introducing more sophisticated modeling techniques.

In recent years, deep learning has been widely applied to visual tasks, and integrating DCF with deep features has become a major direction for improvement. However, directly incorporating deep convolutional features from networks such as VGG [53] or ResNet [34] significantly increases computational cost, thereby compromising real-time performance. ToMP [49] replaces the optimizer module in DCF with a Transformer, directly predicting target model weights through self-attention mechanisms to enhance model flexibility. PrDiMP [18] introduces a probabilistic regression framework, improving DCF's confidence prediction by estimating the conditional probability density of the target state, thereby better modeling annotation noise and task ambiguity. With the increasing deployment of visual tracking in resource-constrained environments such as mobile devices, drones, and intelligent robots, DCF trackers must operate efficiently under limited computational budgets. This calls for the development of compact DCF tracking models through approaches like filter compression, model quantization, and frequency-domain acceleration, enabling low-power, high-speed performance suitable for embedded applications.

2.4 One-Stream Framework

2.4.1 Paradigm Introduction

The One-Stream Framework is a widely adopted neural network architecture in computer vision tasks, commonly used in VOT, action recognition, and video understanding. Its core characteristic lies in processing different inputs through a unified feature extraction pathway, enabling feature sharing and end-to-end optimization. Unlike the Two-stream Framework, the one-stream architecture employs a single input channel during data processing, integrating spatial information, temporal dynamics, target appearance, and background cues within a unified learning pathway. This design simplifies the overall model structure, reduces computational overhead, and facilitates joint optimization in an end-to-end manner. Typically, this framework feeds both the target template and the search region into a shared feature extraction network—often a lightweight CNN or a residual network—to obtain high-level semantic representations. This shared architecture avoids the feature distribution inconsistency issues often encountered in traditional multi-branch structures. In typical Siamese-based one-stream tracking networks, both the template and the search image are encoded by the same network, and a cross-correlation operation is performed in the feature space to establish matching relationships, generating a response map to localize the target. To improve target localization accuracy, some extended versions introduce additional regression and classification branches. These branches, while maintaining the single-stream structure, enable multitask learning for bounding box prediction and category discrimination. This framework is not only compact and computationally efficient but also well-suited for end-to-end training and deployment, making it particularly advantageous for real-time vision applications.

The core idea of the One-Stream framework is to jointly model and learn features from multiple modalities—or, in the case of VOT, from the template and search region—within a single unified network pathway. This architecture fuses the inputs at an early stage and processes them through a shared deep neural network, enabling end-to-end modeling and collaborative understanding of cross-modal or cross-region information. A typical One-Stream network comprises an input fusion module, followed by several convolutional layers, normalization and activation units, pooling operations, and potentially fully connected layers or Transformer-based modules. Depending on the task, the network may adopt 2D or 3D convolutions to simultaneously capture spatial and temporal features.

Unlike traditional Two-Stream frameworks, where features from the template and search region are extracted independently and fused only at later stages, the One-Stream approach allows early-stage semantic alignment and interaction modeling. This continuous joint representation throughout the network enhances the discriminability and adaptability of the learned features. To address modality or input discrepancies, One-Stream models often incorporate auxiliary mechanisms such as modality embeddings, attention modules, or channel weighting strategies—either before or after the input fusion—to facilitate effective integration. Additionally, cross-modal alignment strategies or cooperative attention can be employed to mitigate redundancy and inconsistency, ultimately improving the robustness of the joint representation. This paradigm shift breaks through the limitations of branch independence and delayed relational modeling in traditional tracking pipelines, delivering more efficient and expressive VOT performance.

In terms of algorithmic implementation, the one-stream framework is typically based on a Transformer architecture. Its input consists of a pair of images—template, and search—divided into fixed-size patches that are linearly projected to obtain token representations. Positional encodings are added to preserve spatial ordering. These tokens are then concatenated into a single sequence and passed through multiple stacked Transformer encoder layers. During this process, the self-attention mechanism performs not only local context modeling but also establishes global dependencies between template tokens and search tokens. This enables bidirectional guidance and dynamic matching. From a formal perspective of the attention matrix, this mechanism completes both intra-image and inter-image feature fusion and semantic enhancement in a single pass, thus endowing the network with object awareness throughout, rather than relying solely on later-stage matching or aggregation modules.

Furthermore, to balance model performance and computational overhead, the one-stream framework often introduces an early candidate elimination strategy. This approach dynamically scores search region tokens in intermediate layers based on the similarity distribution extracted from the attention maps, retaining only the top-k tokens with the highest relevance to the target. The remaining tokens are either discarded or replaced with placeholders to maintain structural consistency. This mechanism effectively reduces redundant computation, prevents low-relevance regions from interfering with subsequent modeling, and improves inference speed and resource efficiency. Methodologically, the one-stream framework embodies a holistic design philosophy of "unified encoding—joint modeling—efficient inference". Through early-stage interaction, frequent feature fusion, and dynamic token selection, it significantly enhances target discriminability and tracking robustness. It represents a key trend in visual tracking architecture design, evolving toward structural unification and efficient information interaction.

In VOT, the One-Stream framework varies in algorithmic focus across RGB, thermal infrared, LiDAR, and event-based modalities, with key differences in feature fusion, modality adaptation, and spatiotemporal modeling. For RGB-based One-stream methods, the template and search images are typically concatenated or differenced and input into a deep CNN to extract features, followed by classification and regression branches to localize the target. These methods emphasize appearance modeling and often incorporate attention mechanisms, self-attention modules, or Transformers to enhance contextual understanding and discriminative power, making them well-suited for scenes with rich textures. In thermal-based One-stream tracking, due to the lack of texture and color in infrared images, lightweight networks are used for feature extraction, often with modality-aware modules to improve feature discrimination. In RGB-Thermal fusion, One-stream methods adopt an early-fusion strategy, concatenating multi-modal images for unified encoding and using modality-guided or weighted fusion mechanisms to address cross-modal redundancy and inconsistency. LiDAR-based One-stream tracking addresses the sparsity of point clouds by first converting them into depth or Bird's Eye View (BEV) maps. A unified network is then used to extract spatial structural features, often incorporating geometry-aware modules to support structure matching and target regression. For event-based One-stream tracking, the focus is on temporal modeling. Event data are aggregated over time windows into pseudo-frames or spatiotemporal volumes, then concatenated with the template and fed into the network. Temporal convolutions or attention mechanisms are used to extract dynamic features, enabling temporal alignment. These methods are particularly effective in fast-motion and dynamic environments. Overall, while sharing a unified architecture, One-stream frameworks are adapted across modalities with specialized algorithmic strategies to ensure robust and accurate tracking performance.

2.4.2 *Typical Methods*

Within the One-Stream framework, OSTrack [65] and MixFormer [14] are two representative classical methods. In the following, we provide a systematic overview of their network architectures and key algorithmic components using these methods as examples.

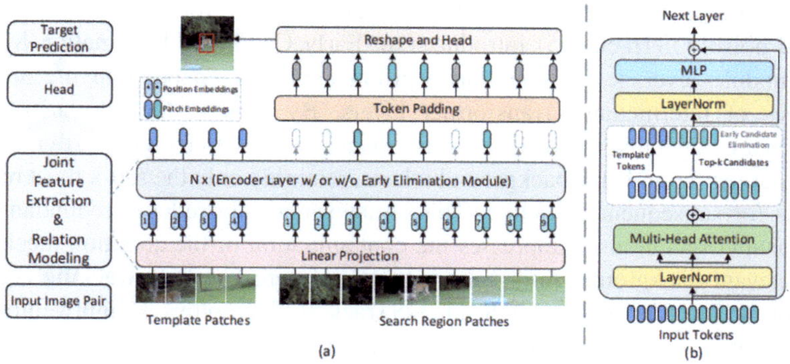

Fig. 2.9 OSTrack Network Architecture Diagram [65]

OSTrack [65] proposes an integrated one-stream VOT network architecture, breaking away from the limitations of traditional two-stream, two-stage frameworks. It achieves joint feature learning and relational modeling between the template image and the search region through a unified Transformer backbone. As illustrated in Fig. 2.9, OSTrack [65] adopts a pre-trained Vision Transformer (ViT) as its backbone to simultaneously process the template and the search region images. During the input stage, the model receives a pair of input images—namely, the target template and the search region from the current frame. These images are first divided into fixed-size patches, which are then linearly projected into a shared embedding space to generate low-dimensional representations. To retain spatial information, independent learnable positional encodings are added to each embedding. After embedding generation, the patch embeddings from both the template and search region are concatenated into a single unified token sequence, which is subsequently passed through a stack of Transformer encoder layers. Each encoder layer within this module consists of standard components, including LayerNorm, self-attention, and a feed-forward neural network.

The core idea of OSTrack [65] lies in introducing a One-Stream architecture, which unifies the traditionally decoupled processes of template and search branch separation and feature extraction and relation modeling into a single, shared feature modeling pipeline. Unlike conventional Two-Stream trackers that extract features from the template and search images independently and later fuse them via an external matching module, OSTrack [65] concatenates both inputs into a unified token sequence from the very beginning. These images are divided into fixed-size patches, linearly projected into a unified token representation, and added with independent 1D positional encodings to retain spatial information. The combined tokens are then passed through a Vision Transformer (ViT) [25] backbone, where stacked Transformer encoder layers perform joint modeling of global spatial dependencies. This enables the template to guide the search region's feature learning in a dynamic and target-aware manner, while inherently integrating relation modeling into the feature extraction stage.

To further enhance inference efficiency and improve robustness against background noise, OSTrack [65] introduces an Early Candidate Elimination Module. This module leverages the similarity priors explicitly encoded in the attention matrices of intermediate Transformer layers. By ranking the similarity scores between the target token and all candidate tokens in the search region, it dynamically filters out low-confidence background tokens, retaining only the top-k most relevant regions for subsequent modeling. This strategy not only reduces redundant computational costs but also suppresses the contamination of the attention mechanism by irrelevant background information, thereby improving both tracking accuracy and speed. For the prediction head, OSTrack [65] employs a lightweight fully convolutional structure to reconstruct and decode the search region tokens. It produces a classification confidence map, positional offsets, and bounding box-size parameters, enabling precise localization of the target via a simple yet effective coordinate regression strategy. Another notable advantage of this approach lies in its strong scalability and training efficiency. By adopting a ViT backbone, OSTrack [65] can directly inherit large-scale pre-trained models (e.g., MAE-pre-trained ViT), which significantly accelerates convergence and enhances generalization ability, particularly demonstrating stronger cross-category discriminative capability in one-shot tracking scenarios.

Key Algorithm Formulas in OSTrack [65]

1. **Interaction formula in the self-attention mechanism:**

$$A = \text{Softmax}\left(\frac{QK^\top}{\sqrt{d_k}}\right) \cdot V \tag{2.31}$$

$$V = \text{Softmax}\left(\frac{[Q_z; Q_x][K_z; K_x]^\top}{\sqrt{d_k}}\right) \cdot [V_z; V_x] \tag{2.32}$$

This formula describes the bidirectional information interaction process between the template z and the search region x in the self-attention mechanism. Specifically, the query (Q), key (K), and value (V) matrices of both are concatenated, and the attention weights are calculated. The model can dynamically fuse the features of the template and the search region.

2. **Similarity calculation of the candidate elimination module:**

$$h_z^i = \text{Softmax}\left(\frac{q_i \cdot [K_z; K_x]^\top}{\sqrt{d}}\right) \cdot V = [w_z^i; w_x^i] \cdot V \tag{2.33}$$

This formula is used to calculate the similarity between the template center and the candidate regions in the search area. By selecting the attention weights at the template center, the template can filter out background candidate regions.

3. **The formula for target localization:**

$$(x, y, w, h) = (x_d + O(0, x_d, y_d),\ y_d + O(1, x_d, y_d),\ S(0, x_d, y_d),\ S(1, x_d, y_d)) \tag{2.34}$$

This formula generates the final target bounding box based on the classification score (P) and regression outputs (O and S) of the search region feature map. Here, O represents the local offset compensation for the discretization error caused by resolution reduction, and S represents the normalized bounding box size. Since the features have been dynamically fused through the One-Stream framework, high-precision results can be obtained directly from the output without the need for additional matching modules.

MixFormer [14] is an end-to-end object tracking framework based on the Transformer architecture, characterized by a clear hierarchical feature processing pipeline. Its core innovation lies in the introduction of the Mixed Attention Module (MAM), which synchronously achieves feature extraction and target information integration within a unified framework. MAM combines both self-attention and cross-attention mechanisms, simultaneously performing feature extraction and interaction between the target template and search region in a single step. Self-attention operates on the token sequences of the target template and searches regions independently, extracting local and global features. Meanwhile, cross-attention facilitates bidirectional information interaction between the target and search region, enhancing target awareness. Additionally, MAM incorporates an asymmetric attention variant, which prunes redundant cross-attention connections between the target and search region to reduce computational costs. This design also supports online multi-template updates, further enhancing tracking adaptability.

The network architecture of MixFormer [14] follows a multi-stage progressive design, composed of several stacked Mixed Attention Modules (MAMs). As shown in Fig. 2.10, the system input consists of the target image (128 × 128) and the search image (320 × 320). Initially, these input images are split into feature blocks of size H × W × 3, including the target template (128 × 128 × 3), and the search region

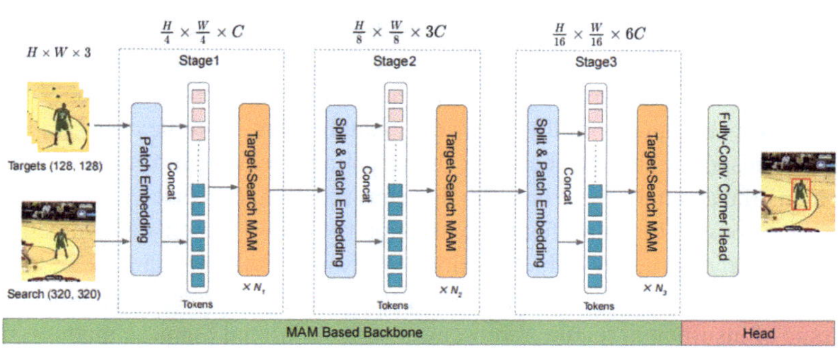

Fig. 2.10 MixFormer Network Architecture Diagram [14]

(320 × 320 × 3), through a shared-weight Patch Embedding layer. The system processes the images through three stages, with each stage employing overlapping convolutional embedding layers that progressively reduce spatial resolution while increasing the channel dimension, mapping from the initial spatial space to token space. In the first stage, the output feature maps of the search region and template have 64 (MixFormer) or 192 (MixFormer-L) channels. The subsequent stages further downsample the feature maps, ultimately outputting feature maps of size 20 × 20 (search region) and 8 × 8 (template) with channel dimensions of 384 and 1024, respectively.

The network architecture adopts a three-stage progressive feature extraction backbone, where each stage integrates the core Target-Search Mixed Attention Module (MAM) to achieve efficient and deep cross-modal feature interaction and fusion. In Stage 1, the input images are first transformed into unified visual token representations through segmentation and reorganization operations. The feature resolution is downsampled to H/4 × W/4, while the channel dimension is expanded to 3C, providing a larger representational capacity for subsequent deep modeling. In Stage 2, the network further compresses the feature resolution to H/8 × W/8 and extends the channel dimension to 6C. Within this stage, the MAM module is applied repeatedly N2 times, strengthening the feature interaction and context aggregation between the target template and the search region. Finally, Stage 3 maintains the same spatial resolution but deepens the fusion at the feature level. The stacked attention mechanisms in this stage enable the modeling of higher-order semantic representations. Notably, the network no longer relies on a separate relation head or additional interaction modules; instead, the interaction between the template and search regions is integrated directly into the backbone, enabling end-to-end deep fusion. To further enhance the model's representational capacity, MixFormer [14] introduces two types of target localization heads. The first is a fully convolutional corner-based head, which predicts spatial heatmaps of the target's bounding box corners and applies expectation computation to regress the final box. The second is a Transformer-style decoder based on learnable query tokens, which directly regresses the bounding box coordinates, achieving a fully sequence-based modeling approach in line with the Transformer framework.

The MAM module, as the core component of the network structure, integrates both self-attention and cross-attention mechanisms. It processes the token streams of the target searches images in parallel and uses channel dimension concatenation to achieve cross-modal information fusion. This module not only effectively enhances the semantic alignment ability between features but also flexibly adjusts the similarity response strength between the target and background information, thus improving the adaptability to complex interference scenarios (such as occlusion and scale variations). The entire feature extraction backbone presents a symmetric parallel dual-stream structure, which retains the advantages of the Siamese architecture in terms of parameter sharing and compact structure. At the same time, it overcomes the issue of insufficient feature representation in traditional Siamese methods due to a lack of deep interaction by introducing high-frequency interaction modules. The data path, indicated by the arrows in the diagram, clearly reveals the layer-by-layer evolution of the information flow from low-level details to high-level semantics, reflecting the effectiveness of the multi-scale progressive learning strategy.

In terms of the design of the localization head, MixFormer [14] introduces two different implementation strategies: the first is a fully convolutional localization head based on corner prediction, which first predicts the probability distribution maps of the top-left and bottom-right corners of the target bounding box using multiple convolutional layers. Then, a weighted expectation method is applied to regress the final bounding box coordinates. The second strategy is a regression-based localization head based on query tokens, where learnable regression tokens are introduced, and the Transformer decoder structure is used to directly regress the bounding box coordinates, achieving an end-to-end pure Transformer target localization solution. The former has the advantages of a simple structure and fast inference, while the latter better aligns with the unified modeling philosophy of Transformer, offering stronger representational power and cross-scale adaptability. To further enhance the model's stability in long-sequence tracking tasks, MixFormer [14] also introduces an online template dynamic update mechanism. This mechanism leverages the confidence prediction module (SPM), which uses the interaction features between the target template and the search region to generate confidence scores for the matching quality of the current frame. Template updates are only allowed when this score exceeds a set threshold, effectively preventing cumulative drift caused by incorrect matches. This mechanism achieves a balance between discriminative and adaptive template updates, significantly improving the model's robustness and long-term consistency in practical applications.

In terms of dynamic adaptation, MixFormer [14] introduces an online template updating strategy by designing a Similarity Prediction Module (SPM) to assess the quality of the current frame's matching. When the predicted matching confidence extracted from the interaction features exceeds a predefined threshold, the system proactively updates the template to prevent long-term drift caused by tracking failures. The strength of MixFormer [14] lies in its end-to-end unified design, which abandons the conventional decoupled CNN+Transformer architecture. Instead, it couples feature extraction and target integration via the Mixed Attention Module (MAM), significantly improving both the discriminative power of features and the efficiency of feature interaction.

Key Algorithmic Formulas in MixFormer [14]

1. **Key-Value Concatenation:** This formula concatenates the keys (k_t, k_s) and values (v_t, v_s) from the template and search region into a unified set of key-value pairs, forming a single-stream input:

$$K = [k_t; k_s], \quad V = [v_t; v_s] \tag{2.35}$$

2. **Mixed Attention Computation:** The template branch (q_t) and the search branch (q_s) share a common set of keys and values (k_m, v_m), enabling bidirectional information exchange through cross-attention. Both self-attention within each sequence and cross-attention across sequences are performed jointly in a single

computation, eliminating the need for separate processing steps:

$$\text{Attention}(q, k_m, v_m) = \text{Softmax}\left(\frac{qk_m^\top}{\sqrt{d}}\right) v_m \tag{2.36}$$

MixFormerV2 [15] proposes a distillation-based model compression paradigm, including dense-to-sparse distillation (transferring knowledge from dense corner head to sparse token prediction) and depth-to-shallow distillation (reducing network depth through progressive layer pruning and intermediate teacher models). This significantly improves tracking efficiency while maintaining high accuracy, making it suitable for real-time deployment on both GPU and CPU platforms. SBT [61] dynamically integrates target template information into the feature extraction process of the search image using a multi-level cross-attention mechanism. It generates instance-discriminative feature representations that are directly used for target localization prediction. It incorporates spatial reduction global attention (SRG) to reduce computational complexity and accelerates convergence using pre-training strategies. SimTrack [7] constructs a unified feature representation by serializing inputs, utilizing self-attention mechanisms to achieve multi-level bidirectional interactions between the template and search region. It introduces a center-cropped window strategy to mitigate information loss during downsampling, thus simplifying the architecture while enhancing tracking accuracy. DropMAE [59] proposes an adaptive spatial attention-dropping strategy. It dynamically suppresses spatial attention weights within frames and strengthens cross-frame temporal attention interaction. This forces the model to learn more discriminative temporal correspondences during the mask reconstruction process, improving the representation learning performance for video object tracking and segmentation tasks. F-BDMTrack [64] introduces a foreground-background distribution modeling-based visual tracking method. Through the FBAL module, it dynamically generates foreground-background agents (FB agents) for both the template and search region. The DA2 module computes distribution similarity between features, rather than appearance similarity, to suppress background interference, thereby enhancing the learning ability of discriminative target features. GRM [28] uses a learnable token partition module (based on Gumbel-Softmax for end-to-end optimization) to adaptively select tokens from the search region that are relevant to the template for cross-modal interaction. Early layers suppress background interference, while later layers degrade into global interactions, thus unifying the advantages of both two-stream and one-stream paradigms. HIPTrack [5] introduces a History Information Pondering (HIP) module, which incorporates fine-grained target masks to introduce historical position information. By combining search region features and encoding historical appearance information, it uses an L2 distance-based similarity aggregation strategy to retrieve multi-frame historical cues from a memory bank. This enhances the feature representation of the current frame via channel concatenation, significantly improving the tracker's robustness to complex scenarios like target deformation and occlusion without requiring fine-tuning of the backbone network.

RFGM [68] designs a correlation attention mechanism that adaptively selects the most relevant historical features from the GR Memory for the current search region. It also uses adaptive token ordering to update the GR Memory at the token level, reducing redundant calculations and suppressing interference from irrelevant information. ROMTrack [6] maintains the purity of the initial target features using an inherent template, performs bidirectional feature matching between the template and search region with mixed templates, and introduces lightweight variation tokens for dynamic modeling of target appearance changes. This enhances the tracker's adaptability to deformations and appearance changes while suppressing background interference. SeqTrack [9] models visual tracking as a sequence generation task. It discretizes boundary box coordinates into discrete tokens and uses an encoder–decoder Transformer to autoregressively predict the sequence, requiring only a single cross-entropy loss. This eliminates the need for complex multitask head networks and redundant loss functions common in traditional trackers. ARTrack [58] models visual tracking as a temporal autoregressive coordinate sequence prediction task. It discretizes trajectory coordinates to build a vocabulary and uses a structured loss function to maximize the sequence likelihood while incorporating Intersection over Union (IoU) loss to enhance spatial correlation supervision. This enables end-to-end motion modeling and unified optimization of visual template matching. VideoTrack [62] extends the traditional Siamese tracking's 2D single-frame matching to spatiotemporal modeling at the video level. It directly encodes temporal context through an end-to-end video Transformer, and a decoupled dual-template mechanism separates static appearance and dynamic state information. ODTrack [67] introduces a video-level tracking framework that redefines target tracking as a temporal token autoregressive propagation task. The autoregressive propagation task compresses discriminative target features into token sequences and iterates them across frames, enabling efficient spatiotemporal modeling without complex online updates. This significantly improves long-term association and motion trajectory modeling capabilities. EVDTrack [52] propagates spatiotemporal tokens across frames and generates explicit multi-scale and spatiotemporal cues, which are directly input into the Transformer encoder for feature interaction. This avoids the "when to update" and "how to update" issues of traditional template update strategies, achieving efficient utilization of spatiotemporal information and real-time inference.

2.4.3 *Summary and Challenges*

The one-stream framework demonstrates significant structural advantages and application potential in the field of VOT. Its core strengths lie in the depth of feature interaction, architectural unification, and inference efficiency. The one-stream design offers modular simplicity and end-to-end trainability. Its unified Transformer backbone reduces architectural complexity and facilitates the integration of large-scale pre-trained models, thereby accelerating convergence and boosting cross-domain generalization performance. A key direction in improving the generalizability and robustness of one-stream architectures is to incorporate identity-aware enhancement mechanisms. Alternatively, adaptive branches can be designed to strengthen individual target representations. Both approaches aim to maintain the structural advantages of a shared backbone while improving target-specific expressiveness.

Furthermore, when extending the one-stream framework to more generalized multi-modal tasks, the challenges become even more complex. Significant asynchrony and heterogeneity often exist between modalities—for instance, image and text inputs differ in temporal scale and spatial resolution, while image and knowledge graph data diverge in terms of semantic abstraction. Such misalignment may lead to information interference, feature confusion, or semantic degradation during early fusion stages, adversely affecting overall modeling performance. At the same time, the scarcity of high-quality, well-aligned multi-modal data resources also constrains the broader application of the one-stream architecture. Although deep fusion typically relies on large-scale datasets with precise semantic alignment across modalities, real-world applications often suffer from modality missingness, sparse annotations, or insufficient supervisory signals, which pose significant challenges to the model's generalization capabilities. Moreover, due to the highly implicit nature of its fusion process, the one-stream framework exhibits limited interpretability. This makes it difficult to explicitly trace the individual contributions of each modality during the decision-making process, raising concerns about model trustworthiness and verifiability, especially in high-risk deployment scenarios.

Future research should focus on the integration of efficient attention mechanisms, such as sparse self-attention, low-rank approximations, sliding-window attention, or importance-aware token selection mechanisms, to reduce computational overhead and enhance inference efficiency. Secondly, future efforts could incorporate explicit confidence modeling, learnable gating structures, or differentiable token selection strategies to construct more robust, controllable, and semantically guided token scheduling mechanisms, thereby improving interaction efficiency and model robustness. To enhance individual target modeling capabilities, future work may explore the integration of identity-preserving modules or plug-and-play lightweight auxiliary branches within the shared encoding framework. In heterogeneous modality alignment and fusion modeling, the one-stream framework, when extended to multi-modal tasks, must address cross-modal asynchrony and representation heterogeneity. Future research may focus on the design of modality-adaptive projection layers, cross-modal alignment loss functions, and semantic bridging modules to improve the accuracy and semantic consistency of modality fusion. To enhance the interpretability and debuggability of such models, it is essential to further develop effective visualization tools for multi-modal systems. In particular, mechanisms for visualizing in-module representations are needed to gain deeper insights into model behavior. These tools should help reveal how each modality responds and contributes semantically during the fusion process, ensuring transparency in the model's decision-making.

References

1. Agarap AF (2018) Deep learning using rectified linear units (relu). arXiv preprint arXiv:1803.08375
2. Bertinetto L, et al (2016) Fully-convolutional Siamese networks for object tracking. In: Proceedings of the Computer vision–ECCV 2016 workshops: Amsterdam, The Netherlands, October 8–10 and 15–16, 2016, part II 14. Springer, Berlin, pp 850–865
3. Bhat G, et al (2019) Learning discriminative model prediction for tracking. In: Proceedings of the IEEE/CVF international conference on computer vision, pp 6182–6191
4. Bolme DS, et al (2010) Visual object tracking using adaptive correlation filters. In: 2010 IEEE computer society conference on computer vision and pattern recognition. IEEE, New York, pp 2544–2550
5. Cai W, Liu Q, Wang Y (2023) Learning historical status prompt for accurate and robust visual tracking. In: CoRR
6. Cai Y, et al (2023) Robust object modeling for visual tracking. In: Proceedings of the IEEE/CVF international conference on computer vision, pp 9589–9600
7. Chen B, et al (2022) Backbone is all your need: a simplified architecture for visual object tracking. In: European conference on computer vision. Springer, Berlin, pp 375–392
8. Chen T, et al (2020) A simple framework for contrastive learning of visual representations. In: International conference on machine learning. PmLR, New York, pp 1597–1607
9. Chen X, et al (2023) Seqtrack: sequence to sequence learning for visual object tracking. In: Proceedings of the IEEE/CVF conference on computer vision and pattern recognition, pp 14572–14581
10. Chen X, et al (2021) Transformer tracking. In: Proceedings of the IEEE/CVF conference on computer vision and pattern recognition, pp 8126–8135
11. Chen Z, et al (2022) SiamBAN: target-aware tracking with Siamese box adaptive network. IEEE Trans Pattern Anal Mach Intell 45(4):5158–5173
12. Cheng S, et al (2021) Learning to filter: Siamese relation network for robust tracking. In: Proceedings of the IEEE/CVF conference on computer vision and pattern recognition, pp 4421–4431
13. Chopra S, Hadsell R, LeCun Y (2005) Learning a similarity metric discriminatively, with application to face verification. In: 2005 IEEE computer society conference on computer vision and pattern recognition (CVPR'05), vol. 1. IEEE, New York, pp 539–546
14. Cui Y, et al (2022) Mixformer: end-to-end tracking with iterative mixed attention. In: Proceedings of the IEEE/CVF conference on computer vision and pattern recognition, pp 13608–13618
15. Cui Y, et al (2023) Mixformerv2: efficient fully transformer tracking. Adv Neural Inf Proces Syst 36:58736–58751
16. Dai J, et al (2017) Deformable convolutional networks. In: Proceedings of the IEEE international conference on computer vision, pp 764–773
17. Dalal N, Triggs B (2005) Histograms of oriented gradients for human detection. In: 2005 IEEE computer society conference on computer vision and pattern recognition (CVPR'05), vol. 1. IEEE, New York, pp 886–893
18. Danelljan M, Gool LV, and Timofte R (2020) Probabilistic regression for visual tracking. In: Proceedings of the IEEE/CVF conference on computer vision and pattern recognition, pp 7183–7192
19. Danelljan M, et al (2014) Adaptive color attributes for real-time visual tracking. In: Proceedings of the IEEE conference on computer vision and pattern recognition, pp 1090–1097
20. Danelljan M, et al (2019) Atom: accurate tracking by overlap maximization. In: Proceedings of the IEEE/CVF conference on computer vision and pattern recognition, pp 4660–4669

21. Danelljan M, et al (2016) Beyond correlation filters: learning continuous convolution operators for visual tracking. In: Proceedings of the Computer Vision–ECCV 2016: 14th European Conference, Amsterdam, The Netherlands, October 11–14, Part V 14. Springer, Berlin, pp 472–488
22. Danelljan M, et al (2016) Discriminative scale space tracking. IEEE Trans Pattern Anal Mach Intell 39(8):1561–1575
23. Danelljan M, et al (2017) ECO: efficient convolution operators for tracking. In: Proceedings of the IEEE conference on computer vision and pattern recognition, pp 6638–6646
24. Danelljan M, et al (2015) Learning spatially regularized correlation filters for visual tracking. In: Proceedings of the IEEE international conference on computer vision, pp 4310–4318
25. Dosovitskiy A, et al (2020) An image is worth 16 $\times$ 16 words: transformers for image recognition at scale. arXiv preprint arXiv:2010.11929
26. Du F, et al (2020) Correlation-guided attention for corner detection based visual tracking. In: Proceedings of the IEEE/CVF conference on computer vision and pattern recognition, pp 6836–6845
27. Fan H, et al (2019) Lasot: a high-quality benchmark for large-scale single object tracking. In: Proceedings of the IEEE/CVF conference on computer vision and pattern recognition, pp 5374–5383
28. Gao S, Zhou C, Zhang J (2023) Generalized relation modeling for transformer tracking. In: Proceedings of the IEEE/CVF conference on computer vision and pattern recognition, pp 18686–18695
29. Glorot X, Bordes A, Bengio Y (2011) Deep sparse rectifier neural networks. In: Proceedings of the fourteenth international conference on artificial intelligence and statistics. JMLR Workshop and Conference Proceedings, pp 315–323
30. Gu J, et al (2018) Recent advances in convolutional neural networks. Pattern Recogn 77:354–377
31. Guo D, et al (2021) Graph attention tracking. In: Proceedings of the IEEE/CVF conference on computer vision and pattern recognition, pp 9543–9552
32. Guo D, et al (2020) SiamCAR: Siamese fully convolutional classification and regression for visual tracking. In: Proceedings of the IEEE/CVF conference on computer vision and pattern recognition, pp 6269–6277
33. He A, et al (2018) A twofold Siamese network for real-time object tracking. In: Proceedings of the IEEE conference on computer vision and pattern recognition, pp 4834–4843
34. He K, et al (2016) Deep residual learning for image recognition. In: Proceedings of the IEEE conference on computer vision and pattern recognition, pp 770–778
35. He K, et al (2015) Delving deep into rectifiers: surpassing human-level performance on imagenet classification. In: Proceedings of the IEEE international conference on computer vision, pp 1026–1034
36. He K, et al (2020) Momentum contrast for unsupervised visual representation learning. In: Proceedings of the IEEE/CVF conference on computer vision and pattern recognition, pp 9729–9738
37. Henriques JF, et al (2012) Exploiting the circulant structure of tracking-by-detection with kernels. In: Proceedings of the Computer Vision–ECCV 2012: 12th European conference on computer vision, Florence, Italy, October 7–13, 2012, Part IV 12. Springer, Berlin, pp 702–715
38. Henriques JF, et al (2014) High-speed tracking with kernelized correlation filters. IEEE Trans Pattern Anal Mach Intell 37(3):583–596
39. Howard AG, et al (2017) Mobilenets: efficient convolutional neural networks for mobile vision applications. arXiv preprint arXiv:1704.04861
40. Galoogahi HK, Fagg A, Lucey S (2017) Learning background-aware correlation filters for visual tracking. In: Proceedings of the IEEE international conference on computer vision, pp 1135–1143
41. Krizhevsky A, Sutskever I, Hinton GE (2012) Imagenet classification with deep convolutional neural networks. Adv Neural Inf Proces Syst 25:1097–1105

42. Lavin A, Gray S (2016) Fast algorithms for convolutional neural networks. In: Proceedings of the IEEE conference on computer vision and pattern recognition, pp 4013–4021
43. Li B, et al (2018) High performance visual tracking with siamese region proposal network. In: Proceedings of the IEEE conference on computer vision and pattern recognition, pp 8971–8980
44. Li Y, Zhu J (2014) A scale adaptive kernel correlation filter tracker with feature integration. In: European conference on computer vision. Springer, Berlin, pp 254–265
45. Li Z, et al (2021) A survey of convolutional neural networks: analysis, applications, and prospects. IEEE Trans Neural Networks Learn Syst 33(12), 6999–7019
46. Lin T-Y, et al (2017) Feature pyramid networks for object detection. In: Proceedings of the IEEE conference on computer vision and pattern recognition, pp 2117–2125
47. Liu Z, et al (2021) Swin transformer: hierarchical vision transformer using shifted windows. In: Proceedings of the IEEE/CVF international conference on computer vision, pp 10012–10022
48. Ma C, et al (2015) Long-term correlation tracking. In: Proceedings of the IEEE conference on computer vision and pattern recognition, pp 5388–5396
49. Mayer C, et al (2022) Transforming model prediction for tracking. In: Proceedings of the IEEE/CVF conference on computer vision and pattern recognition, pp 8731–8740
50. Pouyanfar S, et al (2018) A survey on deep learning: algorithms, techniques, and applications. ACM Comput Surv (CSUR) 51(5):1–36
51. Ren S, et al (2015) Faster R-CNN: towards real-time object detection with region proposal networks. Adv Neural Inf Proces Syst 28:91–99
52. Shi L, et al (2024) Explicit visual prompts for visual object tracking. In: Proceedings of the AAAI conference on artificial intelligence, vol. 38(5), pp 4838–4846
53. Simonyan K, Zisserman A (2014) Very deep convolutional networks for large-scale image recognition. arXiv preprint arXiv:1409.1556
54. Valmadre J, et al (2017) End-to-end representation learning for correlation filter based tracking. In: Proceedings of the IEEE conference on computer vision and pattern recognition, pp 2805–2813
55. Vaswani A, et al (2017) Attention is all you need. Adv Neural Inf Proces Syst 30:5998–6008
56. Wang J, et al (2020) Deep high-resolution representation learning for visual recognition. IEEE Trans Pattern Anal Mach Intell 43(10):3349–3364
57. Wang Q, et al (2019) Fast online object tracking and segmentation: a unifying approach. In: Proceedings of the IEEE/CVF conference on computer vision and pattern recognition, pp 1328–1338
58. Wei X, et al (2023) Autoregressive visual tracking. In: Proceedings of the IEEE/CVF conference on computer vision and pattern recognition, pp 9697–9706
59. Wu Q, et al (2023) Dropmae: masked autoencoders with spatial-attention dropout for tracking tasks. In: Proceedings of the IEEE/CVF conference on computer vision and pattern recognition, pp 14561–14571
60. Xia C, Xiong C, Yu P (2021) Pseudo siamese network for few-shot intent generation. In: Proceedings of the 44th international ACM SIGIR conference on research and development in information retrieval, pp 2005–2009
61. Xie F, et al (2022) Correlation-aware deep tracking. In: Proceedings of the IEEE/CVF conference on computer vision and pattern recognition, pp 8751–8760
62. Xie F, et al (2023) Videotrack: learning to track objects via video transformer. In: Proceedings of the IEEE/CVF conference on computer vision and pattern recognition, pp 22826–22835
63. Xu T, et al (2019) Learning adaptive discriminative correlation filters via temporal consistency preserving spatial feature selection for robust visual object tracking. IEEE Trans Image Process 28(11):5596–5609
64. Yang D, et al (2023) Foreground-background distribution modeling transformer for visual object tracking. In: Proceedings of the IEEE/CVF international conference on computer vision, pp 10117–10127
65. Ye B, et al (2022) Joint feature learning and relation modeling for tracking: a one-stream framework. In: European conference on computer vision. Springer, Berlin, pp 341–357

66. Zhang K, et al (2014) Fast visual tracking via dense spatio-temporal context learning. In: Proceedings of the Computer Vision–ECCV 2014: 13th European Conference, Zurich, Switzerland, September 6–12, 2014, Part V 13. Springer, Berlin, pp 127–141
67. Zheng Y, et al (2024) Odtrack: online dense temporal token learning for visual tracking. In: Proceedings of the AAAI conference on artificial intelligence, vol. 38(7), pp 7588–7596
68. Zhou Y, et al (2018) Efficient correlation tracking via center-biased spatial regularization. IEEE Trans Image Process 27(12):6159–6173
69. Zhu Z, et al (2018) Distractor-aware siamese networks for visual object tracking. In: Proceedings of the European conference on computer vision (ECCV), pp 101–117

Chapter 3
Single-Modal Object Tracking

Abstract This chapter explores single-modal object tracking across four sensing modalities: RGB, thermal infrared, LiDAR, and event cameras. Each modality offers distinct characteristics that influence tracking strategies. RGB-based tracking is the most mature, leveraging rich color information and wide applicability. Thermal infrared tracking excels in low-light conditions and temperature-sensitive applications, while LiDAR provides accurate 3D spatial representations critical for autonomous navigation. Event-based tracking offers high temporal resolution and low latency, ideal for dynamic environments. For each modality, the chapter categorizes mainstream tracking paradigms—such as Siamese networks, correlation filters, one-stream architectures, and motion modeling—and outlines representative algorithms and methods. It also compares their performance across benchmarks and highlights ongoing challenges including occlusion, target deformation, illumination variation, and sensor-specific limitations. The structure of the chapter moves from modality characteristics to technical paradigms, method examples, performance evaluation, and concludes with open problems, providing a comprehensive view of single-modal object tracking. This chapter underscores the importance of modality-specific strategies in advancing robust, efficient tracking systems across diverse real-world applications.

3.1 RGB-Based VOT

3.1.1 Modal Characteristics and Application Scenarios

RGB modality represents images or videos using red, green, and blue color channels. It captures and describes color information through these three components. An RGB sensor implements this modality by converting light into digital signals. The sensor typically includes an array of photodiodes, each corresponding to a pixel. A filter array of red, green, and blue optical filters covers the sensor surface. These filters allow specific wavelengths to pass, capturing the intensity of each color. Equal intensities of all three colors produce white. Zero intensity in all channels results in black.

M. Wang et al., *Visual Object Tracking across Modalities*, Advances in Computer Vision and Pattern Recognition, https://doi.org/10.1007/978-981-95-3664-1_3

The advantages of RGB modality are as follows:

1. **Biomimicry**: Human cone cells are most sensitive to red, green, and blue light. The RGB modality is designed based on this principle. By combining these three colors, it can reproduce the full range of colors seen by the human eye. This ensures that the images and videos it produces closely match real-world scenes.
2. **Universality**: RGB sensors are standard in modern imaging devices. They are widely used in smartphones, digital cameras, and displays. This makes the RGB modality a convenient method for capturing visual information.
3. **Low cost**: RGB sensors are inexpensive and based on mature technology. This makes RGB-based systems highly cost-effective. As a result, RGB modality is widely used in consumer electronics, industrial automation, and medical imaging.
4. **Rich color representation**: The RGB modality can generate millions of colors by combining red, green, and blue light. For example, with 8-bit depth, each channel has 256 levels. Together, they can produce over 16 million colors. This allows for high-quality color reproduction in both natural and artistic scenes.

The above advantages make the RGB modality a key technology in modern visual information processing. It supports progress in consumer electronics, industrial automation, medical imaging, artificial intelligence, and more. Usually, the RGB modality is widely applied in many computer vision tasks:

- Semantic Segmentation [159]: Semantic segmentation assigns each pixel or voxel in an image or video to a specific semantic category. Its goal is to identify the location and distribution of different objects within a scene. This enables fine-grained understanding of visual content. Unlike traditional classification, it focuses on labeling every pixel, such as "road", "pedestrian", or "vehicle".
- Action Recognition [235]: Action recognition detects and classifies dynamic behaviors of people or objects from videos or image sequences. It aims to understand ongoing activities and match them to predefined actions, like "running", "waving", or "jumping". It considers not just appearance but also motion patterns and movement over time. The key is to extract and analyze spatiotemporal information across consecutive frames.
- Hand Pose Estimation [127]: Hand pose estimation identifies and tracks hand posture, motion, and position from image or video input. It aims to capture the 3D pose or motion sequence of the hands. These data are translated into a form that computers can understand. It enables real-time understanding of hand gestures and supports intuitive human–computer interaction.

Additionally, RGB cameras are now widely used in many industries. As a result, many downstream tasks have been included in the application scope of the RGB modality. These include wheat stripe rust spread center detection [39], rice crop monitoring [55], black oat biomass estimation [180], bacterial leaf blight detection [6], non-destructive prediction of wheat chlorophyll [166], pig disease symptom monitoring and health management [160], tissue section image analysis [183], 6D pose estimation [148], autonomous driving [262], and visual object tracking (VOT), which will be detailed in this section.

3.1.2 Inductive Classification and Typical Methods

In VOT, RGB-based tracking methods started early. With rapid advances in hardware and theory, they have attracted growing research attention. Following the classification in Chap. 2, as shown in Fig. 3.1, RGB-based VOT methods can be categorized into the following five major paradigms: Siamese network paradigm, correlation filter paradigm, one-stream network paradigm, large model transfer paradigm, and detection/segmentation-tracking paradigm.

Between 2016 and 2020, the Siamese network and correlation filter paradigms were widely used in VOT. They attracted significant attention from the global academic community. Over time, the Siamese network paradigm has stayed at the forefront. Its strong feature extraction and pattern recognition ensure accurate capture of key targets. This holds even in complex scenarios like deformation, background interference, and lighting changes. As a result, it enables stable and efficient tracking. In contrast, the correlation filter paradigm gained early attention in VOT but declined sharply after 2020. It struggles with fast movements, heavy occlusions, and deformations. These limitations have led to reduced research interest.

Meanwhile, the one-stream and large model transfer paradigms have gained momentum in recent years. This is driven by the rise of the Transformer and related technologies. Transformer's strong parallel computing and long-range dependency capture greatly enhance feature extraction, sequence modeling, and target localization. These improvements offer new solutions to complex VOT challenges.

Next, this section will provide an in-depth analysis of each paradigm, covering its overview, subtypes, research priorities, and key methods.

3.1.2.1 Siamese Network Paradigm

As detailed in Chap. 2, the Siamese network paradigm is a significant deep-learning approach in VOT. Its core lies in constructing a pair of weight-sharing Siamese subnetworks. One receives the target template, and the other takes in candidate images from the search region. They extract features and compute similarity for target localization.

Early methods like SiamFC [10] pioneered the application of Siamese networks in tracking. They transformed tracking into a similarity-matching task between the target template and search frames, featuring a fully convolutional structure without fully connected layers. Cross-correlation operations generate response maps to determine the target location. Subsequent methods such as SiamRPN [96] introduced Region Proposal Networks (RPNs), adding classification and

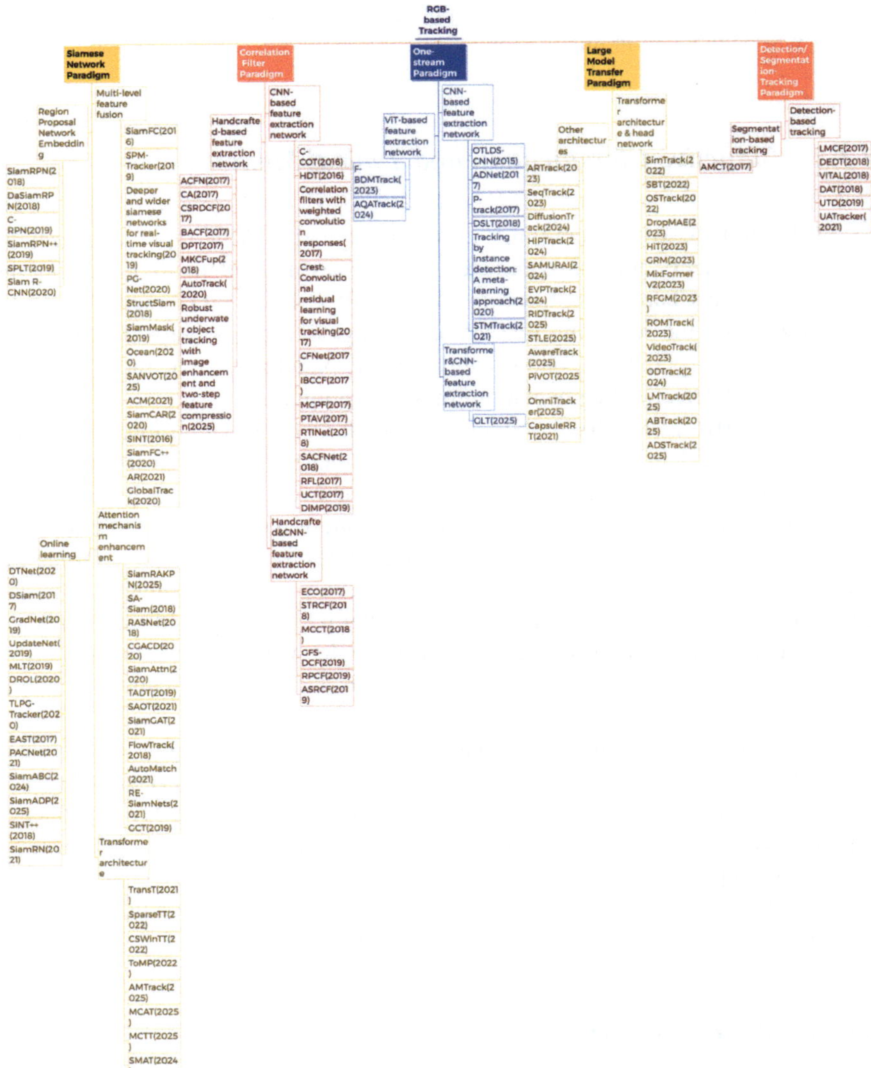

Fig. 3.1 Classification of VOT Method Based on RGB

regression branches to directly predict target-box locations. This eliminated the need for interpolation and multi-scale testing, thereby improving tracking speed and accuracy.

Many other methods have also enhanced feature extraction quality and tracking performance by improving network structures and introducing self-attention mechanisms. With advantages like simple structure, high computational efficiency, and easy online update of target features, the Siamese network paradigm can effectively handle occlusion and target appearance changes. It has achieved remarkable results in VOT and is one of the key research directions in this field. Next, according to the overall network structure design, this paradigm can be further classified into the following five categories: multi-level feature fusion, region proposal network embedding, attention mechanism enhancement, online learning, and Transformer architecture.

I Multi-Level Feature Fusion

The core of these methods lies in using shallow-layer features for detailed info and deep-layer features for semantic info. This enhances the model's ability to distinguish the target and improves robustness. Some methods use multi-branch structures to extract features from different depth levels and fuse them via feature fusion modules like skip connections or attention mechanisms. This increases tracking accuracy without sacrificing computational efficiency. Multi-level feature fusion strategies enable Siamese networks to better handle challenges like target appearance changes, background interference, and scale variations. As a result, they achieve real-time, high-precision visual tracking.

As the pioneering work that introduced the Siamese network paradigm to VOT, SiamFC [10] was the first to use a fully convolutional Siamese network. It achieved real-time and efficient tracking by performing cross-correlation to match features between target templates and search regions. However, its shallow network structure limits feature representation. Subsequent methods have advanced by integrating deeper network architectures, multi-level feature fusion, and refined matching strategies. These enhancements address challenges such as target appearance changes, background clutter, and scale variations, leading to significantly improved tracking accuracy and robustness.

SPM-Tracker [184] adopts a series–parallel matching structure with a two-stage process: a coarse matching phase to ensure robustness and a fine matching phase to enhance distinctiveness. The work in [253] designs padding-free residual modules (CIR units) to build deeper networks, optimizing stride and receptive field to improve localization accuracy. PG-Net [115] gradually narrows the search range via a pixel-to-global matching module, combining a shared correlation architecture to reduce computational costs. StructSiam [252] constructs target structural relationships through local structural pattern detection and message-passing modules, enhancing robustness to deformation and occlusion. SiamMask [191] jointly learns similarity matching, bounding box regression, and segmentation masks, improving tracking and segmentation accuracy with multi-task feature fusion. Ocean [255] employs an anchor-free design for direct target position regression and a feature alignment module to sample target-aware features from predicted bounding boxes. SANVOT [17] initializes the target position with a YOLO-V5 detector and uses a Siamese network to adaptively regress the bounding box scale and aspect ratio. ACM [71] decomposes convolution operations to fuse heterogeneous features and combines bounding box prior information to enhance target-background distinguishability. SiamCAR [65] predicts classification scores and bounding box offsets pixel-wise, avoiding anchor box hyperparameter design to simplify tracking. SINT

[178] uniformly samples multi-scale candidate regions on a circle and selects the most similar region via a Siamese network. SiamFC++ [215] features a multi-branch design for classification, regression, and quality assessment, suppressing low-confidence bounding boxes with quality scores. AR [218] decodes bounding box coordinates with a corner prediction head and refines target localization with auxiliary segmentation masks providing pixel-level supervision. GlobalTrack [84] avoids local assumptions with a global search framework, processing each frame independently with query-guided RPN and RCNN.

II Region Proposal Network Embedding

The Region Proposal Network (RPN) was originally introduced in the field of object detection to generate high-quality candidate regions for subsequent classification and regression tasks. In Siamese networks, RPN is ingeniously embedded to leverage its region proposal advantages. By combining with the similarity measurement capability of Siamese networks, RPN effectively locates target objects within the search region. This integration allows tracking algorithms to maintain high efficiency while significantly enhancing the accuracy and robustness of target localization. The incorporation of RPN into Siamese networks improves candidate region generation. Multi-level features further enhance feature representation. Online updating mechanisms help the tracker adapt to complex scenes and target variations.

As the first tracker to introduce RPNs into the Siamese network framework, SiamRPN ushered in a new era for visual tracking. Siamese networks provide efficient feature extraction. RPNs offer precise region proposals. By combining both, the tracker directly outputs classification scores and bounding box regression in a single step. This offers an efficient and accurate solution for real-time object tracking. Subsequent tracking algorithms have refined RPN structures to better handle complex scenes. These improvements also enhance robustness against similar distractors.

DaSiamRPN [269] explicitly generates semantic distractors as negative samples to train the RPN and uses a distractor-aware module to suppress similar backgrounds online. C-RPN [45] uses cascaded RPN modules to filter out easy negatives step by step. It also employs feature transfer blocks to fuse multilayer features, improving localization accuracy. SiamRPN++ [97] introduces a spatially aware sampling strategy to solve the translation deviation caused by deep network padding and enhances scale adaptability through multilayer feature aggregation. SPLT [217] generates candidate boxes using RPN. An offline-trained deep verifier selects the best proposals, enabling global search for long-term tracking. Siam R-CNN [182] adopts the two-stage detection framework of Faster R-CNN. It compares template and candidate features using a re-detection head. Trajectory dynamic programming is used to enhance long-term robustness.

III Attention Mechanism Enhancement

The core idea of attention-enhanced methods in Siamese networks is to strengthen target feature representation. This leads to improved accuracy and robustness in VOT. Specifically, these methods first extract the features of target templates and search regions through the two branches of the Siamese network. Then, they employ attention mechanisms to capture the contextual relationships between targets and backgrounds. This highlights the distinctiveness of target features and suppresses background interference. In this way, these methods effectively handle complex variations in target appearance, scale, and posture, and greatly improve adaptability to challenges like occlusion and illumination changes during tracking.

SiamRAKPN [93] combines self-attention to capture the internal structure of targets and cross-attention to integrate background context, enhancing distinctiveness through attention mechanisms. SA-Siam [72] introduces a channel attention mechanism in the semantic branch, calculating weights based on activated channels in the target region to better capture discriminative semantic features. RASNet [192] combines general, residual, and channel attention mechanisms, using weighted cross-correlation layers to adaptively adjust feature interactions between templates and search regions. CGACD [43] generates attention weights based on spatial and channel-wise correlations between target templates and search regions. SiamAttn [231] employs a deformable cross-attention module to model global dependencies between targets and search regions. TADT [109] selects the most relevant feature channels via target-aware filters. SAOT [267] uses a saliency region mining module to locate key local features and model their global correlations. SiamGAT [64] establishes part correspondences between target templates and search regions based on graph attention networks. FlowTrack [270] aligns and aggregates historical frame features through a spatiotemporal attention module. AutoMatch [254] automatically selects the best feature fusion operators and optimizes the matching process with binary channel decisions. RE-SiamNets [70] achieve rotation equivariance through channel attention. GCT [50] models spatiotemporal relationships using graph convolutional networks.

IV Online Learning

These methods are based on a Siamese network architecture. This architecture uses a two-branch structure. One branch extracts features from the target template. The other extracts features from the search region. Matching is performed through cross-correlation operations. This allows them to maintain high accuracy and efficiency in complex tracking scenarios. The introduction of an online learning module improves adaptability. It updates feature representations in real time. This helps the model handle changes in the target's appearance and background. As a result, tracking becomes more robust and adaptable. These methods adjust the feature template online or learn target-specific feature spaces. This allows them to deal with deformation, occlusion, and complex backgrounds. At the same time, they maintain real-time performance.

DTNet [251] employs a deep reinforcement learning framework to dynamically switch between a detection tracker and a template tracker, adjusting model weights online for dynamic tracking updates. DSiam [66] updates the target appearance transformation matrix (V) and background suppression matrix (W) online, using FFT for real-time adaptation. GradNet [103] dynamically updates the target template using backpropagation gradients, capturing appearance changes while maintaining discriminability. UpdateNet [246] learns template update strategies via a convolutional network, generating adaptive templates from initial, historical, and predicted templates. MLT [29] generates target-specific weights through meta-learning, updating the appearance model with minimum entropy memory samples. DROL [263] enhances the Siamese network's discriminability with an online classifier, using short-term templates to handle deformations and occlusions. TLPG-Tracker [107] optimizes target model weights online, refining localization with deformable center correlation operations. EAST [81] uses a reinforcement learning agent to dynamically select feature levels, applying deep features for significant appearance changes. PACNet [238] employs a policy network to decide search patterns and an actor-critic network for precise bounding box regression. SiamABC [237] dynamically updates batch normalization statistics to adapt to target distribution shifts during testing. SiamADP [172] detects input blurriness with a motion blur perception module and restores clear features via a deblurring network. SINT++ [197] enhances training data diversity with a positive sample generation network and a hard positive transformation network. SiamRN [28] distinguishes targets from distractors with a meta-learning relationship detector and contrastive training.

V Transformer Architecture

These methods use a twin-branch Siamese network structure. One branch processes the template. The other processes the search region. The network learns similarities and differences between them. At the same time, they introduce the self-attention mechanism and encoder–decoder structure of Transformer architecture to enhance global feature modeling and feature interaction. These methods use self-attention and cross-attention mechanisms. They integrate features from templates and search regions. This results in more discriminative feature representations. This improves tracking accuracy and robustness. Some methods also reduce computational complexity by using lightweight Transformer modules and efficient feature fusion strategies. This allows them to adapt to resource-constrained hardware and meet real-time requirements.

TransT [26] enhances template and search region features with self-attention (ECA) and models their global dependencies via cross-attention (CFA). SparseTT [47] focuses on target-relevant areas with sparse attention and extracts hierarchical features using a Swin Transformer backbone. CSWinTT [170] preserves target integrity with cyclic-shift window attention and addresses boundary distortion with a spatial regularization mask. ToMP [137] predicts target model weights with a Transformer, enabling end-to-end training and inference instead of traditional optimizers. AMTrack [134] models target motion patterns with spatiotemporal-frequency attention and integrates high- and low-frequency information with a hybrid frequency module. MCAT [193] gradually optimizes features with local-to-global progressive attention and captures information at different abstraction levels with multi-scale pixel-level cross-correlation. MCTT [32] suppresses interfering features with multi-head cross-attention and aids positioning by introducing pixel-level spatial information with a segmentation mask head. SMAT [59] simultaneously extracts template and search region features with a separable mixed-attention ViT and refines global context with a decoder self-attention.

3.1.2.2 Correlation Filter Paradigm

As elaborated in Chap. 2, the correlation filtering paradigm for VOT is an efficient tracking method based on signal processing. Its core idea is to transform the target tracking problem into one of filter design and update. By constructing a correlation filter in the frequency domain of images, the filter can produce a significant peak response at the target location. In this paradigm, the initial target image block is first Fourier-transformed to obtain its frequency domain features. Then, the model uses these features and Gaussian labels. The labels have a peak at the target center and lower values around it. A correlation filter is trained with these data. Its parameters are obtained by solving a linear regression problem. When processing later video frames, the model uses the current search region. This region is transformed using the Fourier transform. It is then correlated with the trained filter. This produces a response map. The peak of the map shows the predicted target location.

The correlation filtering paradigm has high computational efficiency. It can ensure tracking accuracy while achieving fast tracking, so it is widely used in VOT tasks and has been deeply researched. Its performance and adaptability are continuously improved by introducing multi-channel features, online updating strategies, etc. According to the feature extraction network design, it can be further divided into three categories: CNN-based, Handcrafted-based, and Handcrafted&CNN-based.

I CNN-Based Feature Extraction Network

The core advantage of these methods lies in CNN's ability to automatically learn highly discriminative feature representations, which significantly enhances tracking accuracy and robustness. Unlike traditional handcrafted features, CNN-based feature extraction networks can adapt to complex target appearance changes, such as illumination, scale, pose variations, and partial occlusions. Moreover, these methods often incorporate online learning strategies, enabling real-time model updates and further improving the ability to adapt to target appearance changes.

Commonly used CNNs include VGGNet (e.g., VGG-16 and VGG-19). Their deep and simple architecture provides rich feature levels. AlexNet, an early deep CNN, has achieved breakthroughs in image classification and is widely used for feature extraction. ResNet uses residual learning. It solves the gradient disappearance problem in deep networks. This makes it suitable for tracking tasks that need deep features. GoogLeNet is known for its Inception modules. It increases network depth and width without extra computation. It is used for multi-scale feature extraction.

C-COT [34] fuses multi-resolution features in the continuous spatial domain. It optimizes a continuous confidence map via implicit interpolation for sub-pixel localization, addressing feature misalignment. HDT [155] constructs weak trackers using VGG multilayer features. It improves tracking by dynamically weighting the fusion of deep semantic and shallow detail features with an improved Hedge algorithm. The method in [76] aggregates responses via feature pyramid weighted fusion. It combines continuous domain joint optimization and sparse update strategies to boost multi-scale adaptability. The work in [168] learns multilayer feature weights end to end. It also designs a continuous-scale optimization layer. This replaces traditional discrete search. As a result, the method improves robustness against target deformation. CFNet [181] integrates correlation filters into a neural network architecture, enabling end-to-end template matching through a differentiable correlation layer. IBCCF [98] uses a boundary-center two-branch filter to capture aspect ratio changes. It enhances complementarity with near-orthogonal constraints and optimizes via ADMM. MCPF [250] guides particle sampling with multitask correlation filters. It reduces computational complexity by fusing responses with weighted aggregation. PTAV [44] separates real-time tracking and periodic verification in a dual-thread collaborative architecture. It dynamically triggers error correction mechanisms to balance efficiency and accuracy. RTINet [224] combines two networks. One extracts deep features. The other is an unfolded ADMM update network. It turns the iterative process into trainable feedforward computation. SACFNet [248] aligns target regions by predicting spatial transformation parameters with deformable convolutions. It optimizes the synergy of spatial alignment and filtering modules through three-stage training. RFL [221] designs a convolutional LSTM network to generate target filters offline. Combined with S-CNN features, it achieves response prediction without fine-tuning. UCT [271] unifies a convolutional architecture to learn filter parameters. Based on the PNR criterion, it dynamically controls model updates to suppress noise. DiMP [11] jointly performs target classification and bounding box regression. Its model prediction network works with discriminative loss optimization. Together, they produce strong discriminative filters.

II Handcrafted-Based Feature Extraction Network

These methods do not rely on deep learning models to automatically learn features. Instead, they use handcrafted feature extraction networks. These networks capture key visual information of target objects. This approach enables efficient object tracking. Moreover, these methods add spatial and temporal regularization to feature extraction. They also combine contextual information. These strategies improve the robustness and accuracy of tracking.

Researchers often use the following handcrafted features to describe target objects. HOG (Histogram of Oriented Gradients) features capture orientation histograms of image gradients. They effectively represent object edges and textures. This makes them suitable for contour and shape description. Color names features encode image pixels using color labels. This helps distinguish objects with different colors. It works well in scenarios where color features are prominent. Color histogram features count the distribution of color components. They are sensitive to changes in color distribution. This makes them suitable for scenes with little color variation. Haar-like features use simple rectangular templates to calculate local contrast and edge information, offering simplicity and speed, and are fitting for real-time tracking. Local Binary Patterns (LBP) features generate binary patterns by comparing pixel gray values with those of neighboring pixels, showing robustness to illumination changes. Additionally, multi-channel feature fusion technology is widely applied. It integrates multiple features, such as HOG and color features, to form a multi-channel feature representation. This approach leverages the strengths of different features to enhance the comprehensiveness and robustness of the target description.

ACFN [30] dynamically selects optimal tracking modules via an attention network and combines active module validation scores for efficient adaptive tracking. CA [142] introduces contextual patches around the target as hard negative samples in correlation filter training, enhancing background discrimination through regularization. CSRDCF [129] proposes spatial-channel dual reliability mechanisms to adjust filter support regions and feature channel weights, improving tracking accuracy for irregular targets. BACF [91] trains correlation filters with dense background negative samples, breaking the traditional circular translation sample limitation to suppress background noise. DPT [174] uses level set segmentation for deformable blocks and coordinates block-wise tracking results with a shape-preserving kernel-based correlation filter, enabling precise estimation of non-rigid object contours. MKCFup [177] reduces interference between kernels by optimizing the upper bound of a multi-kernel objective function, maintaining the advantages of multi-feature fusion at 150FPS. AutoTrack [110] balances local response focus and global update strategies with spatiotemporal adaptive regularization, optimized for real-time tracking in UAV scenarios. The method in [101] is designed for underwater environments, featuring image enhancement and two-step PCA feature compression. It combines a channel reliability weight update mechanism to handle underwater target deformation and degradation.

III Handcrafted & CNN-Based Feature Extraction network

Handcrafted features, like HOG and Color Names, are celebrated for their computational efficiency and robustness to target appearance changes. Their simplicity enables real-time tracking. Meanwhile, CNN-derived features capture deep semantic information. They also detect complex visual patterns. This ensures strong tracking performance in complex scenes and during major appearance changes. By merging these two types of features, correlation filter algorithms maintain fast tracking speed. They also improve the accuracy and robustness of target localization. This fusion allows the algorithms to deliver better results across a variety of challenging visual tracking situations.

ECO [36] compresses model parameters by decomposing convolution operations. It balances efficiency and overfitting risk through GMM sample space compression and a conservative margin update strategy. STRCF [99] integrates temporal–spatial regularization constraints into the SRDCF framework. It enables fast single-sample optimization using ADMM. MCCT [190] combines multi-expert DCF trackers. It dynamically selects reliable tracking results via robustness evaluation and adaptive update strategies. GFS-DCF [214] combines channel-spatial group feature selection with low-rank constraints. It highlights discriminative features and enhances temporal smoothness. RPCF [175] embeds ROI pooling into correlation filter learning. It improves deformation resistance with efficient Fourier-domain solutions. ASRCF [33] designs adaptive spatial regularization weights. It achieves high-precision real-time tracking with a position-scale dual CF model and ADMM optimization.

3.1.2.3 One-Stream Paradigm

As elaborated in Chap. 2, the one-stream paradigm for VOT is an efficient and succinct method that plays a crucial role in VOT tasks. This paradigm uses a single feature extraction process. It handles both the target template and the search region. This design enables continuous tracking of the target. It treats the target template and search region as a unified input and performs feature encoding through a single feature extraction network. Subsequently, these features are utilized to predict the target's state, such as its position and size. The one-stream paradigm enables efficient processing. It significantly reduces computational costs. It also improves processing speed. These advantages make it suitable for real-time tracking. Furthermore, based on the design of the feature extraction network, this paradigm can be further classified into three categories: CNN-based, VisionTransformer-based, and Transformer&CNN-based.

I CNN-Based Feature Extraction Network

Section 3.1.2.2 has already introduced CNN-based feature extraction networks, so it will not be repeated here. Next, the focus will be on specific tracking methods directly.

OTLDS-CNN [77] extracts deep features with a pre-trained CNN and combines online SVM for discriminative modeling. It generates target saliency maps via backpropagation to enhance foreground–background distinction. Pixel-level localization is achieved by fusing saliency maps and motion information through sequential Bayesian filtering. SVM and generative models are updated online to handle appearance changes, forming a hybrid tracking framework. ADNet [236] constructs an action decision network optimized by supervised pre-training and reinforcement learning strategy gradients. It drives dynamic action sequences to approximate target positions and uses a "state–action" interaction mechanism. Actions are input as image patches and historical action vectors, with outputs as action probability distributions for closed-loop control. Online fine-tuning adapts to target deformations and occlusions. P-track [176] models tracking as a partially observable Markov decision process. It uses deep Q-networks to learn multi-modal decision strategies. Confidence heatmaps and appearance models form the state representations. Reinforcement learning helps the tracker balance accuracy and efficiency. This works well under computational constraints. DSLT [128] is based on VGG-16 with a residual feature fusion network. It combines responses from conv4_3 and conv5_3 to enhance multi-scale representations. The proposed contraction loss function focuses on hard samples to improve regression accuracy. A scale pyramid strategy enables end-to-end joint learning for target localization and size estimation. The method in [185] converts target tracking into a meta-learning-driven instance detection task. It pre-trains a detector using MAML for fast adaptation to new targets. Fine-tuning in the initial frame achieves detector domain adaptation, and model parameters are updated online with incremental samples. This overcomes traditional detector limitations on fixed categories and enables efficient instance-level tracking in few-shot scenarios. STMTrack [48] constructs a spatiotemporal memory network to adaptively retrieve multi-frame historical information. Pixel-level similarity calculations generate accurate bounding boxes.

II ViT-Based Feature Extraction Network

This one-stream paradigm employing ViT-based feature extraction networks has unique features. It abandons the traditional two-stream network model. That model processes templates and search regions separately. Instead, it uses a unified architecture for joint feature extraction. This allows efficient information interaction and fusion. ViT has strong global feature modeling capability. It captures long-range dependencies between image patches. This helps handle target posture changes. It also deals well with partial occlusion and complex background interference. Compared to CNNs, ViT, and its variants can perform parallel feature extraction. It adjust window sizes and downsampling strategies. This reduces computational complexity. At the same time, it preserves feature quality. As a result, the method meets the speed requirements of real-time tracking.

F-BDMTrack [220] comprises two key modules: the Fore-Background Agent Learning (FBAL) module and the Distribution-Aware Attention (DA2) module. The FBAL module initializes foreground-background agents. These agents extract foreground and background information from the template and search region. This process guides the subsequent feature interaction. The DA2 module calculates the similarity between each image patch feature and the foreground–background agents. It uses this similarity as attention scores for target-aware feature interaction, suppressing incorrect interactions between foreground and background. This process extracts more discriminative features.

AQATrack [210] introduces a set of learnable autoregressive queries to capture the instantaneous changes in target appearance within a sliding window. A novel attention mechanism in the decoder generates new queries in the current frame through the interaction of existing queries. Combining the initial target template with these learned queries, a spatiotemporal information fusion module (STM) is designed. The STM effectively integrates the target's static appearance and instantaneous changes, enabling more robust target tracking.

III Transformer&CNN-Based Feature Extraction Network

This kind of network combines the strong local feature extraction of CNN and the excellent global feature modeling of Transformer. It keeps CNN's precise grasp of local details like target shape, edges, and texture. Also, it uses Transformer's self-attention to integrate global context. This comprehensive approach allows a better understanding of the target and background relationship in video frames.

GLT [257] uses a dual-branch feature processing architecture. The template branch employs dynamic depthwise convolution to extract local detail features. By using dynamic convolution kernels, it simulates self-attention mechanisms to adjust channel weights and enhance the capture of fine-grained information like shape and texture. The search branch features a residual Transformer structure. Cross-layer residual connections are applied. They preserve accumulated features from multiple attention layers. This helps strengthen global context modeling. The features from both branches are first processed by a fusion module. This module computes their similarity. The results are then passed into a classification-regression prediction head. Finally, the system outputs the target state. A key-frame-driven template update strategy is introduced to address the issue of long-term appearance drift in tracking.

3.1.2.4 Large Model Transfer Paradigm

The VOT paradigm of large-model transfer aims to optimize tracking performance by leveraging the general-purpose feature-extraction capabilities of pre-trained models. These models learn rich visual patterns from large-scale datasets. Parameter transfer to VOT networks reduces reliance on task-specific data. A common tactic is freezing some or all pre-trained model parameters and fine-tuning only specific tracking network modules. For instance, a pre-trained CNN or ViT can be used as a feature extractor. Add lightweight tracking modules like prediction heads or similarity-measurement modules behind it. The backbone network parameters are frozen. Only the new parts are trained for tracking tasks. This retains the pre-trained model's strong feature representation. It also reduces training complexity and lowers the risk of overfitting.

Knowledge distillation is another effective transfer method. It transfers knowledge from a large pre-trained "teacher" model to a smaller "student" tracking model to boost performance. Teacher models have high computational cost. However, they can generate high-quality features. They also produce accurate response maps or intermediate representations. The student model imitates the teacher's outputs or intermediate features. This allows it to inherit strong feature expression. At the same time, it maintains high efficiency. For example, the student can learn to distinguish between target and background features or mimic the teacher's target motion state predictions.

Additionally, parameter-efficient fine-tuning introduces a few trainable parameters (like adapter layers or prompt vectors) to lightly adjust the pre-trained large model. These parameters act as task-specific "modulators", guiding the model's general features to meet VOT needs. This enhances performance without significantly increasing the model's parameter count.

Next, according to the overall network structure design, this paradigm can be further categorized into two types: Transformer-based architecture & head network and other architectures.

I Transformer Architecture & Head Network

Methods in this category use a Transformer as the backbone. Its self-attention mechanism extracts and models features. These features come from input templates and search regions. This helps capture the relationship between the target and the background. In head network design, a lightweight structure is usually adopted. It directly predicts target positions or classifications. These predictions are based on features extracted by the Transformer. This design avoids the computational burden of traditional complex heads. This design keeps the Transformer's strong feature extraction ability. It uses a simplified head network to improve efficiency. The overall model becomes faster and more effective. As a result, it achieves excellent performance in visual tracking tasks.

SimTrack [20] splits input images into patches. These patches are fed into a Transformer backbone. The model enables cross-sample feature interaction. It also uses a corner predictor for accurate target localization. SBT [208] uses a single stream to fuse template and search features, enhancing target feature expression through self- and cross-attention mechanisms. OSTrack [225] employs a ViT to process template-search pairs, achieving end-to-end classification and regression with self-attention interaction. DropMAE [204] is based on a masked autoencoding pretraining framework. It uses an adaptive spatial attention dropout strategy to strengthen temporal clue learning. HiT [90] builds a hierarchical Transformer with bridging modules to fuse multi-scale features. Dual-image position encoding enhances cross-modal interaction. GRM [51] proposes an adaptive token partition mechanism. It employs attention masking and Gumbel-Softmax to dynamically switch between single- and dual-stream interaction modes, alleviating target-background confusion. MixFormerV2 [31] introduces mixed prediction tokens to capture cross-domain correlations. Combined with deep-shallow distillation compression, it enables efficient inference. RFGM [265] constructs a global representation of memory and a correlation attention mechanism. It dynamically filters historical features and compresses redundant computations. Multi-branch predictions for scores, offsets, and sizes enable accurate localization. ROMTrack [15] uses dual-template streams. These streams separate original and interaction features. Variant tokens are introduced to encode deformation contexts. This design improves tracking robustness. VideoTrack [209] encodes spatiotemporal features with triplet blocks. Its decoupled dual-template mechanism separates static and dynamic appearance cues to improve computational efficiency. ODTrack [261] compresses inter-frame localization information with online token propagation. Associating video temporal clues via linguistic-pattern attention enables efficient long-term tracking. LMTrack [211] screens high-quality spatiotemporal features through token context memory. A unidirectional attention mechanism maintains reference feature consistency. ABTrack [222] dynamically determines ViT block skipping and local token pruning to balance computational efficiency and feature integrity. ADSTrack [200] dynamically samples search region tokens based on correlation scores. Auxiliary tokens optimize background interference suppression.

II Other Architectures

Most of the approaches in this category are based on the Transformer architecture, which has been adapted and optimized for the task of visual tracking. Some methods use an autoregressive framework. They treat tracking as a sequence generation task. An encoder–decoder structure is applied to model trajectories. Spatial–temporal prompts help capture the evolution of these sequences. Others introduce a restricted token interaction module and asymmetric attention mechanisms on top of the Transformer. This approach limits the interaction area between the template and the search image. It also uses a separate distillation region strategy. This strategy helps optimize the prediction head's output.

Some architectures use an explicit visual prompt framework. They propagate information through spatiotemporal tokens. These tokens generate visual prompts. The prompts help the model reason about the current frame. Some architectures introduce diffusion models in a novel way. They use point set representations to describe targets. Multilayer diffusion decoders are applied. These decoders gradually refine target position estimates. These designs show deep understanding and innovation in visual tracking. They aim to enhance tracking performance. They also seek to simplify model structures. Another goal is to improve adaptability to complex scenarios.

ARTrack [202] uses an autoregressive coordinate sequence model. It combines an encoder–decoder architecture with historical states to generate spatiotemporal cues and predict target trajectories step by step. SeqTrack [25] converts tracking into a bounding box sequence generation task. Using a causal Transformer for autoregressive decoding of discrete coordinate tokens achieves end-to-end prediction. DiffusionTrack [207] models the diffusion process from noise to target. It uses point-set representations and multi-step refinement to gradually approach the real bounding box. HIPTrack [13] generates a masked feature library via a historical hint encoder. Attention-driven historical feature aggregation is adaptive. SAMURAI [219] is a zero-shot visual tracking framework based on SAM 2. It uses motion modeling and motion-aware memory selection to improve tracking accuracy and robustness in complex scenes without extra training. EVPTrack [164] uses explicit spatiotemporal tokens to propagate historical information. Multi-scale cues enhance adaptation to target deformation. RIDTrack [122] adopts a restricted token interaction module to filter key areas. Asymmetric attention combined with logic distillation optimizes localization accuracy. STLE [241] uses SAM to generate initial masks. It applies a temporal-location enhanced Transformer for segmentation. It also integrates online motion inference. This design improves tracking accuracy. It effectively handles pixel-level object tracking. AwareTrack [240] builds a conscious template self-attention interaction framework. The Siamese network adaptively optimizes template update intervals. PiVOT [24] enhances generic VOT by using a visual prompting mechanism with CLIP, effectively removing distractions and improving accuracy. OmniTracker [187] is based on Swin Transformer and builds a reference-guided feature enhancement module. It integrates deformable DETR with contrastive learning to achieve end-to-end detection-tracking association. CapsuleRRT [133] models part–whole relationships through a spatiotemporal capsule network. Prior routing and knowledge distillation enhance computational efficiency.

3.1.2.5 Detection/Segmentation-Tracking Paradigm

In the field of object tracking, the formation and development of detection-tracking and segmentation-tracking paradigms reflect a natural evolutionary process. Target detection algorithms are capable of accurately identifying the positions and categories of objects within images. These algorithms are applied to each frame of a video. They analyze the target's position changes across frames. This allows them to infer motion trajectories. As a result, they achieve object tracking. Over time, the integration of detection algorithms with time-series analysis methods has given rise to the detection-tracking paradigm. This approach relies on the continuity of detection results to track objects.

The segmentation-tracking paradigm stems from segmentation algorithms. Instance segmentation algorithms can precisely separate objects from the background, offering pixel-level contours and shape details. The accurate position and boundary data from these algorithms are useful for tracking. Changes in segmentation masks across frames are analyzed. These changes reveal the target's movement. They also indicate state transitions. This enables stable tracking. The shift from segmentation to tracking is a natural progression. This leads to the segmentation-tracking paradigm. It uses rich details from segmentation algorithms. These details help improve tracking accuracy. They also enhance robustness.

I Detection-Based Tracking

LMCF [188] combines structured SVM with a correlation filter framework. It speeds up optimization with recurrent feature maps and suppresses model drift through multi-modal detection and high-confidence updates. DEDT [138] maintains a diverse classifier committee based on ensemble learning. It enhances long-term tracking stability by dynamically screening training data with artificial sample generation and QBC strategies. VITAL [169] introduces adversarial generation masks to randomly drop features and strengthen positive sample diversity. Its high-order loss function suppresses interference from simple negative samples. DAT [150] generates attention maps via backpropagation as a regularizer, guiding the classifier to focus on target-robust regions. UTD [83] designs a target-guided module to integrate template features. Combined with MAML meta-learning, it enables few-shot rapid adaptation. UATracker [264] uses uncertainty estimation to filter high-quality samples for model updates. It enhances occlusion resistance with mixed feature enhancement and deformable convolution.

II Segmentation-Based Tracking

AMCT [226] constructs an absorbing Markov chain graph model based on superpixels. It learns similarity weights between superpixels using Support Vector Regression (SVR) and estimates the target segmentation area by calculating absorption times. A global appearance model based on color histograms is used. It filters connected regions as foreground. An "absorbing vertex" mechanism is also introduced. It treats background superpixels as absorbing states. Together, these methods address non-rigid target tracking. The solution combines bottom-up segmentation with top-down model matching.

3.1.2.6 Typical Methods

I Siamese Network Paradigm Typical Method: SMAT [59]

As shown in Fig. 3.2, SMAT targets resource-limited scenarios. It reduces the high computational cost of transformer-based trackers. To achieve this, it uses a separable mixed-attention ViT backbone. It also adopts a separable self-attention transformer head. These components work in concert to achieve efficient feature extraction, information fusion, and target state estimation.

Backbone Network Implementation Details

As shown in Fig. 3.3, the ViT backbone uses mixed attention. It first concatenates features from the target template and search region. Then, it applies a separable mixed attention mechanism. This allows simultaneous modeling of global (self) and cross (between template and search region) attention. Fewer transformer blocks are needed compared to separate self and cross-attention computation, reducing latency.

The mixed-attention ViT backbone takes two inputs: the target template Z_{in} and the search region X_{in}. Initially, CNN-based inverted residual (IR) blocks process these inputs to generate local features and reduce spatial dimensions. In the ViT block, after 3×3 and 1×1 CNN layers with weight sharing, the channel number of Z and X is projected from C to d. The output is tokenized and concatenated to form tokens for mixed-attention learning, enabling joint processing of template and search region features. Finally, 1×1 convolutional filters $(qkv - proj)$ produce the query (Q, a $k \times 1$ vector) and the key ($\mathcal{K}$) and value ($\mathcal{V}$, both $k \times d$ matrices).

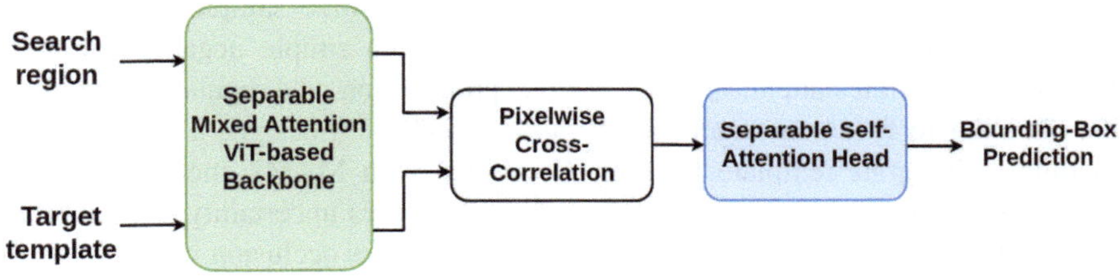

Fig. 3.2 The main architecture diagram of SMAT [59]

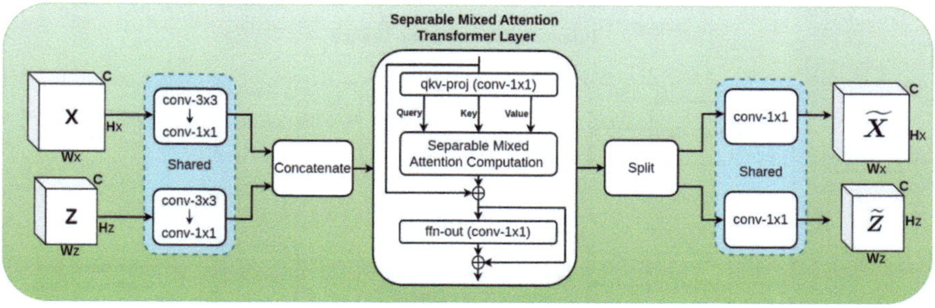

Fig. 3.3 Separable Mixed Attention ViT block [59]

Then, perform the context vector and mixed-attention output calculations separately:

$$\mathcal{A} = \sum_k \tilde{Q} \odot \mathcal{K} \tag{3.1}$$

Here, $\tilde{Q}$ is the extended query vector obtained by applying the softmax operation to Q and broadcasting it column-wise. $\mathcal{K}$ is the key matrix, and $\odot$ denotes element-wise multiplication. $\sum_k$ indicates summation across the rows.

$$\mathcal{M} = \tilde{\mathcal{A}} \odot \text{ReLU}(\mathcal{V}) \tag{3.2}$$

Here, $\tilde{\mathcal{A}}$ is the extended context vector obtained by broadcasting the context vector $\mathcal{A}$ row-wise, $\mathcal{V}$ is the value matrix, and ReLU is the activation function.

Finally, the mixed-attention output feature map is split into template and search region features, each with d channels. A 1×1 convolutional filter reprojects the split features' channels from d back to C, generating the final output features $\tilde{Z}$ and $\tilde{X}$.

Head Network Implementation Details

As shown in Fig. 3.4, the separable self-attention prediction head integrates CNN and transformer layers. It leverages CNNs to model local relationships and transformer layers to explicitly model long-range global interactions. This combination of local and global context modeling enhances the tracker's performance in complex scenarios like drastic target shape changes and severe occlusion.

First, the fusion feature encoding F is obtained by performing pixel-wise cross-correlation ($PWCorr$) on the processed template (Z) and search region (X) features:

$$F = PWCorr(\tilde{X}, \tilde{Z}) \tag{3.3}$$

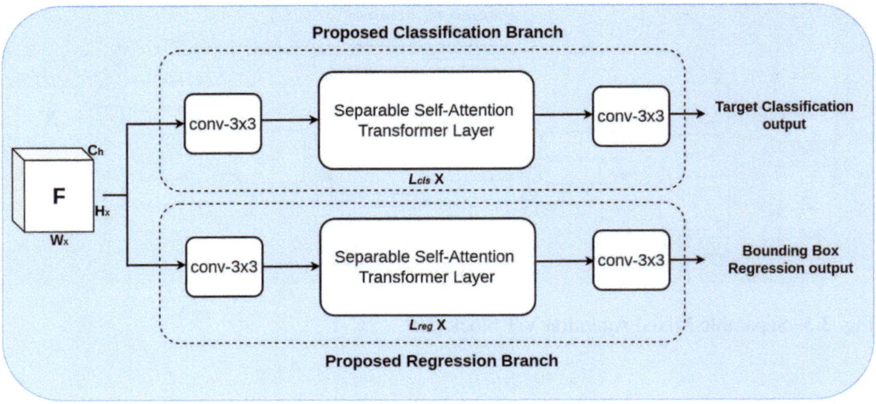

Fig. 3.4 Separable Self-Attention Transformer-based Predictor Head [59]

Then, a 1×1 convolutional filter is applied to the fused feature encoding F to convert the channel number to C_h. The processed F is then input into the classification and regression branches. The classification branch predicts a score map for the target, while the regression branch predicts local offsets and normalized bounding-box sizes. Each branch includes several separable self-attention layers. These layers capture long-range dependencies. Self-attention helps the model understand complex scenes more effectively.

Loss Function Implementation Details

Finally, the output of the classification branch uses the weighted focal loss, and the output of the regression branch uses the ℓ_1 loss and generalized IoU loss. The total loss function is defined as:

$$\mathcal{L}_{total} = \mathcal{L}_{focal} + \lambda_{\ell 1} \cdot \mathcal{L}_{\ell 1} + \lambda_{IoU} \cdot \mathcal{L}_{IoU} \tag{3.4}$$

Here, $\mathcal{L}_{focal}$, $\mathcal{L}_{\ell 1}$, and $\mathcal{L}_{IoU}$ denote the focal loss, ℓ_1 loss, and generalized IoU loss functions, respectively. $\lambda_{\ell 1}$ and λ_{IoU} are hyperparameters that control the relative contribution of each loss function.

II Correlation Filter Paradigm Typical Method: [101]

As shown in Fig. 3.5, underwater tracking faces several challenges. These include visual distortion, color cast, and low visibility. To address these issues, [101] proposed a new tracking framework. It is based on correlation filters. This framework integrates several key components: underwater image enhancement to improve visual quality, a two-step feature compression mechanism to reduce data dimensionality and retain important information, correlation filter learning for accurate target localization, and scale estimation and model updating to adapt to changes in target size and dynamics.

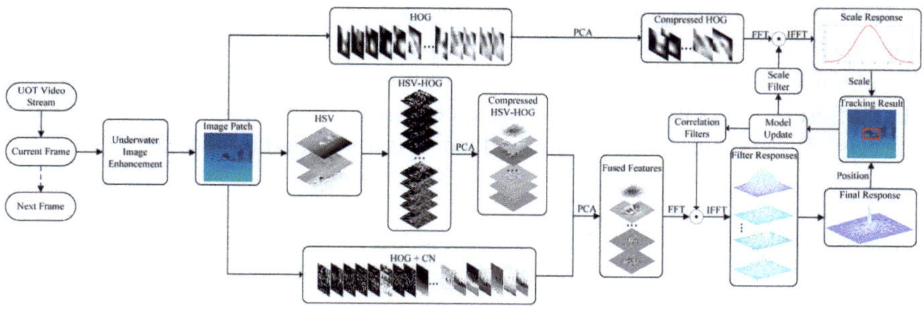

Fig. 3.5 The main architecture diagram of [101]

Image Enhancement Implementation Details

Specifically, the image enhancement process comprises several key steps. Initially, the red light channel is corrected due to its significant absorption underwater. By adjusting the average values of the red and green channels using specific formulas, the image's natural colors are restored. Next, the blue light channel is compensated for, as it is often lost due to underwater scattering. This compensation helps recover the blue in the image. Additionally, image sharpening is crucial, where linear normalization and Gaussian filtering enhance the image's clarity and highlight edges and details. Finally, multiple input images are fused into a high-quality tracking image through weighted blending of Laplacian and Gaussian pyramids.

Two-Step Feature Compression Mechanism Implementation Details

The two-step feature compression mechanism is designed to handle underwater low-visibility conditions. By reducing feature dimensions, it cuts computation costs and retains key information. Firstly, images are transformed to the HSV color space, and HSV-HOG features are extracted and compressed via PCA to form 60-dimensional compressed HSV-HOG features. Then, these are fused with HOG and CN features from the original image. APCA is applied again to reduce the dimensions of the fused features, resulting in a final 40-dimensional feature. This process effectively reduces feature dimensions and enhances target localization accuracy through multi-feature integration.

Correlation Filter Learning Implementation Details

CSR-DCF is an advanced correlation filter method for target tracking. It enhances robustness and accuracy by incorporating channel and spatial reliability:

Given a set of channel features $f = \{f_d\}_{d=1:N_d}$ and corresponding target templates (filters) $h = \{h_d\}_{d=1:N_d}$, where $f_d \in \mathbb{R}^{d_w \times d_h}$ and $h_d \in \mathbb{R}^{d_w \times d_h}$, the target position x is estimated by maximizing the probability $p(x|h)$:

$$p(x|h) = \sum_{d=1}^{N_d} p(x|f_d)p(f_d), \tag{3.5}$$

Here, $p(x|f_d) = [f_d * h_d](x)$ is the convolution result. It combines the feature map and the learned template at position x. $p(f_d)$ is the prior probability. It reflects the reliability of each channel. The optimal filter minimizes the squared difference. This difference is between channel-wise correlation outputs and the target output $g \in \mathbb{R}^{d_w \times d_h}$. A regularization term is added. It is controlled by the parameter λ.

$$\arg_h \min \sum_{d=1}^{N_d} \|f_d * h_d - g\|^2 + \lambda \sum_{d=1}^{N_d} \|h_d\|^2 . \tag{3.6}$$

This problem can be solved in the frequency domain via Parseval's theorem. Specifically, let $\hat{a} = \text{vec}(\mathcal{F}[a])$ represent the Fourier transform of vector a, with $\mathcal{F}$ denoting the Fourier transform operation. By applying complex conjugate gradient solving to each channel, a closed-form solution can be derived. However, this solution may have boundary defects.

CSR-DCF solves the input shift uniformity and pixel reliability issues by adding a spatial reliability map m. This map m identifies pixels to exclude during filter learning, applying the constraint $h \equiv m \odot h$, where $\odot$ denotes element-wise multiplication. This makes the above optimization problem not directly solvable in closed form. CSR-DCF addresses this by introducing dual variable h_c and the constraint $h_c - m \odot h \equiv 0$, thus building an augmented Lagrangian function.

$$\begin{aligned} \mathcal{L}(\hat{h}_c, h, \hat{I}|m) = \left\| \hat{h}_c^H \text{diag}(\hat{f}) - \hat{g} \right\| + \frac{\lambda}{2} \|h_m\|^2 \\ + \left[\hat{I}^H (\hat{h}_c - \hat{h}_m) + \overline{\hat{I}^H (\hat{h}_c - \hat{h}_m)} \right] + \mu \left\| \hat{h}_c - \hat{h}_m \right\|^2 , \end{aligned} \tag{3.7}$$

Here, $\hat{I}$ is a complex Lagrange multiplier. By applying the ADMM algorithm, the augmented Lagrangian function can be solved to obtain a closed-form solution.

During correlation filter learning, a feature weighting method is introduced. It uses the PSR (Peak-to-Sidelobe Ratio) metric. This method improves the efficiency of multi-channel feature fusion. PSR, defined as the ratio of the correlation output's peak to the average sidelobe level, assesses each channel's response quality. By assigning reliability weights to each channel, the target location can be predicted more accurately.

Given the channel features $f = \{f_d\}_{d=1:N_d}$ and target templates $h = \{h_d\}_{d=1:N_d}$, the final response $\tilde{g}$ is obtained by weighting each channel's response.

$$\tilde{g} = \sum_{d=1}^{N_d} f_d * h_d \cdot \tilde{w}, \tag{3.8}$$

Here, $\tilde{w}_d$ is the channel reliability weight, calculated as follows:

$$\tilde{w} = \frac{\text{PSR} - \min(\text{PSR})}{\max(\text{PSR}) - \min(\text{PSR})}. \tag{3.9}$$

Scale Estimation and Model Updating Implementation Details
In the scale estimation and model updating phase, the method first conducts precise scale estimation using a 1D scale correlation filter. Specifically, it defines a scale space S, extracts HOG features related to scale from it, and applies PCA for dimensionality reduction to boost algorithm efficiency. Then, it generates correlation response maps with a scale filter. Through correlation operations, multiple response maps are obtained. The optimal scale factor corresponding to the maximum value in these maps is determined to estimate the target's scale.

After locating the target's position and scale, the system updates the correlation filters. These include both translation and scale filters. It also updates the channel reliability weights. These updates help the tracker adapt to dynamic environmental changes. This update combines the previous frame's filter with the current frame's filter estimation results. Here, the learning rate η plays a key role, balancing filter learning and enabling the tracking system to promptly adapt to target changes. Similarly, channel weights are updated based on current frame estimations. These operations enhance the tracking system's robustness against target position and scale variations.

III One-Stream Paradigm Typical Method: AQATrack [210]
As shown in Fig. 3.6, existing VOT algorithms face two main issues. They lack deep spatiotemporal information mining and rely too much on hand-crafted components. To address these problems, a new method called AQATrack is proposed. It uses a self-regressive, query-based adaptive tracking approach. It introduces learnable autoregressive queries. These queries help capture rapid changes in the target's appearance. A novel attention mechanism is designed to generate new queries for the current frame. The method also uses a Spatio-Temporal Information Fusion Module (STM). This module combines the target's static appearance with its instantaneous changes. This approach effectively solves two major problems. It improves the use of spatiotemporal information. It also reduces reliance on complex

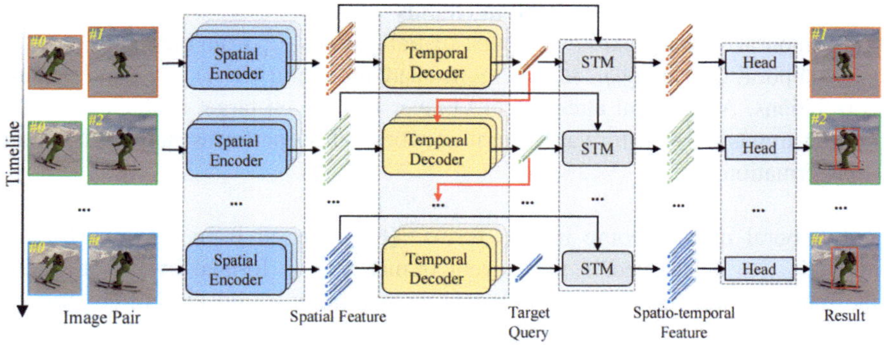

Fig. 3.6 The main architecture diagram of AQATrack [210]

hand-designed components. As a result, tracking robustness and accuracy are significantly enhanced:

Spatial Encoder Implementation Details

The spatial encoder in AQATrack extracts spatial features from the input template and search region images. These features provide essential representations. They support the temporal decoder and the spatiotemporal information fusion module. AQATrack adopts HiViT as its spatial encoder. This encoder uses a 4×4 patch size for patch embedding, which helps in effectively preserving the spatial information of images.

Let the template image be $Z \in \mathbb{R}^{H_z \times W_z \times 3}$ and the search region image be $X \in \mathbb{R}^{H_x \times W_x \times 3}$. After processing by the spatial encoder, the template feature vector $f_z \in \mathbb{R}^{N_z \times D}$ and the search feature vector $f_x \in \mathbb{R}^{N_x \times D}$ are obtained, where $N_z = \frac{H_z W_z}{16^2}$, $N_x = \frac{H_x W_x}{16^2}$, and $D = 512$.

The template and search features are concatenated as $f_{zx}^0 = \text{Concat}(f_z, f_x)$ and then passed through N encoder layers for feature learning. Each encoder layer comprises a multi-head attention mechanism and a feed-forward neural network. The specific operations are defined by the following equations:

$$f_{zx}^n = \text{Encoder}(f_{zx}^{n-1}), \quad n = 1 \dots N \tag{3.10}$$

Here, Encoder denotes the operation of each encoder layer. Finally, the feature f_{zx}^N undergoes layer normalization (LN) processing:

$$f_{zx}^N = \text{LN}(f_{zx}^N) \tag{3.11}$$

Through these steps, the spatial encoder extracts spatial features from the input images. These features offer strong representations. They support the tracking tasks that follow.

Temporal Decoder Implementation Details

The temporal decoder takes two inputs: spatial features f_{zx}^N from the spatial encoder and temporal queries Q_{all}. It uses multi-head attention (MHA) to capture feature correlations. A temporal attention (TA) layer focuses on target state changes and motion trends. A feed-forward neural network (FFN) performs nonlinear feature transformation.

Temporal queries come in two forms: Q_{pre}, passed from previous frames, and Q_{cur}, to be learned in the current frame. Q_{all} is formed by concatenating the temporal queries from the previous $m - 1$ frames with the current frame's query, expressed as $Q_{all} = \text{Concat}(Q_{pre}, Q_{cur})$. This allows the current frame to reference multiple previous frames. It supports a sliding-window method to learn temporal information.

To better capture target changes over time, a temporal attention mechanism is designed. Its calculation is similar to multi-head attention. However, it differs in how the inputs are generated. The query Q comes from the current frame's query Q_{cur}. The key K and value V come from Q_{all}, which includes both historical and current queries. The specific formula is as follows:

$$\text{MultiHead}(Q, K, V) = \text{Concat}(H_1, \ldots, H_{n_h})W^O \tag{3.12}$$

Here,

$$H_i = \text{Attention}\left(QW_i^Q, KW_i^K, VW_i^V\right) \tag{3.13}$$

$$\text{Attention}(q, k, v) = \text{Softmax}\left(\frac{qk^\top}{\sqrt{d_k}}\right)v \tag{3.14}$$

Here, $W_i^Q \in \mathbb{R}^{d_m \times d_k}$, $W_i^K \in \mathbb{R}^{d_m \times d_k}$, $W_i^V \in \mathbb{R}^{d_m \times d_v}$ and $W^O \in \mathbb{R}^{n_h \times d_m}$ are learnable parameters.

In this way, the temporal decoder captures the target's temporal changes. This improves the model's tracking ability. It also makes the model more robust in complex scenarios.

STM Implementation Details

The core concept of the STM is to integrate spatial and temporal information effectively without extra parameters. This is done by computing the similarity between the search region's spatial features f_x^N and the temporal information f_{temporal} using a dot product. Mathematically, the similarity matrix S is expressed as:

$$S = f_x^N \odot f_{\text{temporal}}^\top \tag{3.15}$$

Here, $\odot$ denotes the dot product operation. $f_x^N \in \mathbb{R}^{N_x \times D}$ represents the spatial features of the search region, $f_{\text{temporal}} \in \mathbb{R}^{N_{\text{temporal}} \times D}$ represents the temporal information, and $D = 512$ is the feature dimension. Each element of the similarity matrix $S \in \mathbb{R}^{N_x \times N_{\text{temporal}}}$ shows the similarity between spatial features and temporal information. This helps highlight the potential location of the target.

Then, the similarity matrix S is used to enhance the expressiveness of the spatial features f_x^N. By performing element-wise multiplication between S and f_x^N, temporal information is incorporated into the spatial features. This simple fusion method combines static and dynamic appearance information. It helps the model adapt to target appearance changes. It also improves robustness in cluttered scenes.

AQATrack uses the STM to combine spatial and temporal information. This fusion requires no extra parameters. It improves both tracking accuracy and robustness.

Loss Function Implementation Details

The loss function of AQATrack consists of three main components: Focal Loss for classification, GIoU Loss for bounding box regression, and L1 Loss for offset prediction. By combining these losses, the model is comprehensively supervised to enhance tracking accuracy and robustness.

$$L = L_{\mathrm{cls}} + \lambda_{\mathrm{IoU}} L_{\mathrm{IoU}} + \lambda_{L1} L_{\mathrm{L1}} \tag{3.16}$$

Here,

L_{cls} is the classification loss, L_{IoU} is the Generalized IoU (GIoU) Loss, L_{L1} is the L1 Loss. λ_{IoU} and λ_{L1} are regularization parameters set to 2 and 5, respectively, to balance the weights of different losses.

3.1.3 RGB-Based VOT Methods Performance Comparison

Table 3.1 shows the performance of RGB-based trackers from 2019 to 2025. It compares various methods across multiple datasets. These include OTB100, UAV123, NFS30, VOT-2018, TrackingNet, LaSOT, and GOT-10k. It lists the publication year of each tracker. The table also includes key evaluation metrics. These include success rate (SR), precision rate (PR), and expected accuracy (EAO). It also shows success rate (A), precision rate (R), tracking speed (FPS), and other relevant indicators on the datasets. The comparison shows steady progress in RGB-based VOT methods. Performance has improved across most datasets. The gains are especially clear on challenging ones like LaSOT and GOT-10k. Success and precision rates have improved significantly. This indicates stronger robustness in complex scenarios. These include illumination changes, occlusion, target deformation, and cluttered backgrounds. Additionally, the increase in tracking speed (FPS) reflects enhanced real-time applicability. The table also highlights the performance differences and strengths of various tracking paradigms, such as Siamese network and correlation filter approaches.

3.1.4 Summary and Challenges

This section focuses on single-modal object tracking methods based on RGB modality. RGB modality, offering biomimicry, universality, low cost, and rich color representation, is key for many computer vision tasks. In VOT, RGB-based methods are widely studied due to their early start and the fast progress of related hardware and theory. These methods are divided into five paradigms: Siamese network, correlation filter, one-stream, large model transfer, and detection/segmentation tracking. Each paradigm is further classified based on specific network design details. The Siamese network paradigm remains state of the art. It offers powerful

Table 3.1 Comparison of the performance of different RGB-based trackers from 2019 to 2025

Trackers	Publication	OTB100		UAV123		NFS30		VOT-2018			TrackingNet		LaSOT			GOT-10k			FPS
		SR(%)	PR(%)	SR(%)	PR(%)	SR(%)	PR(%)	EAO	A	R	SR(%)	PR(%)	SR(%)	PR(%)	NPR(%)	AO(%)	S0.5(%)	S0.75(%)	
ASRCF [1]	CVPR2019	69.2	92.2	–	–	–	–	–	–	–	–	–	35.9	33.7	–	–	–	–	28
C-RPN [45]	CVPR2019	66.3	–	–	–	–	–	–	–	–	66.9	61.9	45.9	–	–	–	–	–	36
RPCF [175]	CVPR2019	69.6	93.2	–	–	–	–	–	–	–	–	–	–	–	–	–	–	–	5
SiamMask [191]	CVPR2019	–	–	–	–	–	–	0.387	0.642	0.276	–	–	–	–	–	–	–	–	55
SiamRPN++ [97]	CVPR2019	69.6	91.4	61.3	80.7	–	–	–	–	–	73.3	69.4	49.6	56.9	–	–	–	–	35
SPM-Tracker [184]	CVPR2019	68.7	89.9	–	–	–	–	–	–	–	–	–	–	–	–	–	–	–	120
TADT [109]	CVPR2019	66.0	86.6	–	–	–	–	–	–	–	–	–	–	–	–	–	–	–	33.7
DiMP [11]	ICCV2019	68.4	–	65.4	–	62.0	–	0.440	0.597	0.153	74.0	68.7	56.9	–	–	61.1	71.7	49.2	43
GradNet [103]	ICCV2019	63.9	86.1	–	–	–	–	–	–	–	–	–	36.5	35.1	–	–	–	–	80
GFS-DCF [214]	ICCV2019	69.3	93.2	–	–	–	-=	0.397	0.511	0.143	60.90	56.57	–	–	–	–	–	–	8
MLT [29]	ICCV2019	61.1	–	–	–	–	–	–	–	–	–	–	36.8	–	–	–	–	–	48
SPLT [217]	ICCV2019	–	–	–	–	–	–	–	–	–	–	–	–	–	–	–	–	–	25.7
UpdateNet [246]	ICCV2019	–	–	–	–	–	–	0.481	0.610	0.206	67.7	62.5	47.5	56.0	–	–	–	–	–
DROL [263]	AAAI2020	71.5	93.7	62.5	85.5	–	–	0.481	0.616	–	74.6	70.8	53.7	62.4	–	–	–	–	60
GlobalTrack [84]	AAAI2020	–	–	–	–	–	–	–	–	–	70.4	65.6	52.1	52.7	59.9	–	–	–	6
SiamFC++ [215]	AAAI2020	68.3	–	–	–	–	–	0.426	0.587	0.183	75.4	70.5	54.4	–	–	59.5	69.5	47.9	160
DTNet [251]	NeurIPS2020	66.0	89.1	64.9	83.1	–	–	0.418	0.604	0.197	73.7	69.8	51.6	57.9	–	–	–	–	36
TLPG-Tracker [107]	IJCAI2020	–	69.8	–	–	–	–	0.459	0.606	0.149	75.8	70.6	58.1	–	–	62.9	73.5	–	36
PG-Net [115]	ECCV2020	69.1	89.2	–	–	–	–	0.447	0.618	0.192	–	–	53.1	60.5	–	–	–	–	42
Ocean [255]	ECCV2020	68.4	92.0	–	–	–	–	0.489	0.598	0.117	–	–	56.0	56.6	–	61.1	72.1	–	58
MAML [185]	CVPR2020	71.2	92.6	–	–	–	–	0.452	0.635	0.159	75.7	–	52.3	–	–	–	–	–	40
SiamCAR [65]	CVPR2020	69.6	91.0	63.1	83.3	59.4	–	0.452	0.597	0.178	–	–	51.4	–	59.8	–	–	–	52.27
Siam R-CNN [182]	CVPR2020	70.1	89.1	64.9	83.4	63.9	–	0.408	0.624	0.220	81.2	80.0	64.8	–	72.2	64.9	–	–	15.2
SiamAttn [231]	CVPR2020	71.2	92.6	65	84.5	–	–	0.47	0.63	0.16	75.2	–	56.0	–	64.8	–	–	–	45
AutoMatch [254]	ICCV2021	71.4	92.6	–	–	–	–	–	–	–	76.0	72.6	–	59.9	–	65.2	76.6	54.3	50
SAOT [267]	ICCV2021	71.4	92.6	–	–	65.6	77.8	0.501	–	–	–	–	61.6	–	70.8	64.0	74.9	–	29

(continued)

Table 3.1 (continued)

Trackers	Publication	OTB100		UAV123		NFS30		VOT-2018			TrackingNet		LaSOT			GOT-10k			FPS
		SR(%)	PR(%)	SR(%)	PR(%)	SR(%)	PR(%)	EAO	A	R	SR(%)	PR(%)	SR(%)	PR(%)	NPR(%)	AO(%)	S0.5(%)	S0.75(%)	
TransT[26]	CVPR2021	69.4	–	69.1	–	65.7	–	–	–	–	81.4	80.3	64.9	69.0	73.8	72.3	82.4	68.2	50
STMTrack[48]	CVPR2021	71.9	–	64.7	–	–	–	0.447	0.590	0.159	80.3	76.7	60.6	63.3	69.3	64.2	73.7	57.9	37
SiamRN[28]	CVPR2021	70.1	93.1	64.3	86.1	–	–	0.470	0.595	0.131	–	–	52.7	53.1	–	–	–	–	–
SiamGAT[64]	CVPR2021	71.0	–	64.6	84.3	–	–	–	–	–	–	–	53.9	53.0	63.3	62.7	74.3	48.8	165
UATracker[264]	AAAI2021	70.5	–	67.6	87.9	–	–	0.458	0.614	0.159	–	–	59.4	–	68.0	–	–	–	45
CSWinTT[170]	CVPR2022	–	–	70.5	90.3	–	–	–	–	–	81.9	79.5	66.2	70.9	75.2	69.4	78.9	65.4	12
MixFormer[31]	CVPR2022	–	–	70.4	91.8	–	–	–	–	–	83.9	83.1	70.1	76.3	79.9	71.2	80.0	67.8	25
SBT[208]	CVPR2022	71.9	92.4	–	–	–	–	–	–	–	–	–	66.7	71.1	–	70.4	80.8	64.7	62
ToMP[137]	CVPR2022	70.1	–	69.0	–	66.9	–	–	–	–	81.5	78.9	69.5	73.5	79.2	–	–	–	24.8
SparseTT[47]	IJCAI2022	70.4	–	70.4	–	–	–	–	–	–	81.7	79.5	–	70.1	74.8	69.3	79.1	63.8	39
OSTrack[225]	ECCV2022	–	–	70.7	–	–	–	–	–	–	83.9	83.2	71.1	77.6	81.1	73.7	83.2	70.8	105.4
SimTrack[20]	ECCV2022	–	–	71.2	91.6	–	–	–	–	–	83.4	87.4	70.5	76.2	79.7	69.8	78.8	66.0	40
F-BDMTrack [220]	ICCV2023	69.9	–	70.9	–	67.3	–	–	–	–	84.5	84.0	72.0	77.7	81.5	75.4	84.3	72.9	–
ROMTrack [15]	ICCV2023	71.4	–	–	–	68.0	–	–	–	–	83.6	82.7	69.3	75.6	78.8	72.9	82.9	70.2	62
RFGM [265]	NeurIPS2023	71.5	–	68.5	–	–	–	–	–	–	84.7	83.6	70.3	76.4	82.0	74.1	84.6	71.8	71
MixFormerV2 [31]	NeurIPS2023	–	–	69.9	92.1	–	–	–	–	–	83.4	81.6	70.6	76.2	80.8	73.9	–	–	165
ARTrack [202]	CVPR2023	–	–	71.2	–	67.9	–	–	–	–	85.6	86.0	73.1	80.3	82.2	78.5	87.4	77.8	26
DropMAE [204]	CVPR2023	69.6	–	–	–	–	–	–	–	–	84.1	–	71.8	78.1	81.8	75.9	86.8	72.0	–
GRM [51]	CVPR2023	–	–	70.2	–	65.6	–	–	–	–	84.0	83.3	69.9	75.8	79.3	73.4	82.9	70.4	45
SeqTrack [25]	CVPR2023	–	–	68.5	–	66.2	–	–	–	–	85.5	85.8	72.5	79.3	81.5	74.8	81.9	72.2	5
VideoTrack [209]	CVPR2023	–	–	69.7	89.9	–	–	–	–	–	83.8	83.1	70.2	76.2	–	72.9	81.9	69.8	–
HIPTrack [13]	CVPR2024	71.0	–	70.5	–	68.1	–	–	–	–	84.5	83.8	72.7	79.5	82.9	77.4	88.0	74.5	–
ODTrack [261]	AAAI2024	72.4	–	–	–	–	–	–	–	–	86.1	86.7	74.0	82.3	84.2	78.2	87.2	77.3	32
EVPTrack [164]	AAAI2024	–	–	70.9	–	–	–	–	–	–	84.4	–	72.7	80.3	82.9	76.6	86.7	73.9	28
ARPTrack [114]	CVPR2025	–	–	–	–	–	–	–	–	–	86.6	87.4	74.2	81.7	83.4	81.5	90.6	80.5	91
OmniTrack [186]	TPAMI2025	–	–	–	–	–	–	–	–	–	83.4	82.3	69.1	77.3	75.4	–	–	–	41.7
SPMTrack [14]	CVPR2025	–	–	–	–	–	–	–	–	–	87.3	88.1	77.4	85.0	86.6	81.0	89.2	82.3	–

capabilities in feature learning and representation. The correlation filter paradigm has seen declining attention after 2020 due to limitations in handling complex situations. Meanwhile, the one-stream and large model transfer paradigms are gaining new potential with the rise of Transformer technology.

However, the VOT applications based on RGB modality still face the following challenges:

1. **Lack of depth information**: RGB modality only provides 2D information such as target color and texture and cannot directly obtain depth information. In complex scenes, tracking becomes more difficult. This is especially true when multiple targets look similar. Challenges also arise when the target and background share similar depths. Without depth information, algorithms struggle to judge front-back positions. They also find it hard to understand the target's spatial layout. This increases the risk of misjudgment and tracking failure.
2. **Illumination variation**: Under different lighting conditions, the appearance, color, and texture of the target may change significantly. For example, lighting changes can affect the target. Moving from bright to dark areas may darken the target. Moving from dark to bright may make it brighter. The target may also show color distortion. This increases the difficulty of accurately tracking the target.
3. **Occlusion**: During the tracking process, the target may be partially or completely occluded by other objects. Occlusion hides part of the target. In some cases, the target becomes completely invisible for a while. This makes it hard for tracking algorithms to locate the true position. It also affects their ability to assess the target's current state.
4. **Target deformation**: The target may deform during movement, such as changes in human posture or bending of an animal's body. Deformation alters the appearance and shape characteristics of the target, posing a challenge for tracking algorithms. It requires tracking algorithms to adapt to non-rigid changes in the target and accurately capture its new features.
5. **Background clutter**: Complex backgrounds contain a large number of objects and information similar to or interfering with the target. This can cause tracking algorithms to mistakenly regard background objects as the target, leading to tracking drift or failure. Background clutter can resemble the target. It may share similar colors, textures, or shapes. This makes it harder to distinguish the target from its surroundings.
6. **Scale variation**: The target may appear to change in scale due to different distances or viewing angles, becoming larger or smaller in the image. Tracking algorithms need to adapt in real time to scale changes and accurately detect and locate the target at different scales. Otherwise, mismatches in scale may cause the target to be lost or result in incorrect tracking outcomes.
7. **Fast motion**: When the target moves at a fast speed, the displacement between consecutive frames is large, leading to rapid changes in the target's appearance and possible motion blur. Tracking algorithms may fail to capture updates in time. This delays the detection of the target's latest position and state. As a

result, tracking becomes more difficult. It also raises the need for better real-time performance.

8. **Similar distractors**: There may be other objects in the scene that have a similar appearance to the target. These can interfere with the tracking algorithm's judgment, making it difficult to distinguish the true target from distractors. This can result in tracking errors or frequent switching of the tracked target.

3.2 Thermal Infrared VOT

3.2.1 Modal Characteristics and Application Scenarios

The thermal infrared modality is an imaging technology based on the thermal radiation characteristics of objects. It generates images or data by detecting the infrared radiation emitted from the surface of objects. The core principle is that objects with a temperature above absolute zero ($-273.15°C$) radiate electromagnetic waves. The thermal infrared modality captures the far-infrared band. This band ranges from 8 to 14 microns. Some systems also include the mid-wave band. The mid-wave infrared ranges from 3 to 5 microns.

The advantages [234] of the thermal infrared modality are as follows:

1. **Lighting Independence**: Thermal infrared tasks are not affected by lighting changes. These cameras detect radiation, not visible light. They capture differences in radiation intensity. This radiation comes from objects with different temperatures. This allows them to operate in total darkness, rain, or fog.
2. **Temperature Sensitivity**: All targets emit thermal radiation within a specific wavelength range, with different temperatures leading to different radiation intensities. Thermal infrared cameras capture radiation differences. They provide temperature information for each object. These cameras can detect heat-emitting targets. Examples include humans and vehicle engines. They work even when backgrounds look similar.
3. **Material Penetration**: Thermal infrared radiation can penetrate some non-transparent materials like light smoke, thin fog, and clothing. This helps in complex environments by enabling more comprehensive target information acquisition and enhancing task robustness and accuracy. For instance, in scenarios with light smoke or thin fog, the thermal infrared modality can still function normally.

Thermal infrared modality is widely used in multiple fields [75]. In security and surveillance, it excels in nighttime or low-light conditions. It can be combined with deep learning methods like SSD or YOLOv3 to enhance detection accuracy. It's also applicable for detecting and tracking animals in laboratories, vehicles, and objects in street scenes. In the medical field, thermal infrared imaging can be used to detect diseases like breast cancer and diabetic foot ulcers. For instance, LeNet-based

CNN can classify breast infrared images, and C-DCNN can segment breast regions to identify tumors. It also aids in behavior detection by learning spatiotemporal representations of human actions through deep learning methods combining convolutional and LSTM layers. In industrial inspection, thermal imaging detects heat anomalies. It helps identify issues in electrical equipment and machines. It can also predict component locations and types. Additionally, it monitors structural health. For example, it reveals surface or internal damage in steel bridges. In agriculture and remote sensing, it can assess plant water stress. It's also useful in photovoltaic inspection to detect and classify defects in photovoltaic modules and in material analysis to identify defects in composite materials. These applications highlight the strong potential of thermal infrared modality in detection and analysis under special lighting conditions.

In addition [234], it can be applied in assisted driving and UAV reconnaissance. In assisted driving, its lighting independence allows effective target tracking in nighttime, rain, or fog, ensuring safety. In UAV reconnaissance, it penetrates night and light fog for long-range monitoring of personnel and vehicles, obtaining enemy dynamics.

3.2.2 Inductive Classification and Typical Methods

As VOT based on the RGB modality develops, thermal infrared-based methods have drawn academic attention. Building on Chap. 2's classification paradigm, as shown in Fig. 3.7, thermal infrared-based VOT methods can be divided into the following four major paradigms: Siamese network, correlation filter, one-stream network, and detection-tracking paradigms.

Research on VOT using thermal infrared modality has followed the development trend of RGB modality. In the early stages, Siamese networks and correlation filters were key focus areas, producing many fundamental results. However, thermal infrared tracking has seen slower progress. It has attracted less research interest in recent years. In contrast, RGB-based methods have grown rapidly in other areas.

Next, this section will provide an in-depth analysis of each paradigm, covering its overview, subtypes, research priorities, and key methods.

3.2.2.1 Siamese Network Paradigm

Chapter 2 and Sect. 3.1.2.1 have introduced the Siamese network paradigm, so no further elaboration will be provided here. Subsequently, based on the overall network structure design, this paradigm can be further categorized into five types: multi-level feature fusion, attention-mechanism enhancement, online learning, spatiotemporal context learning, and cross-modal learning.

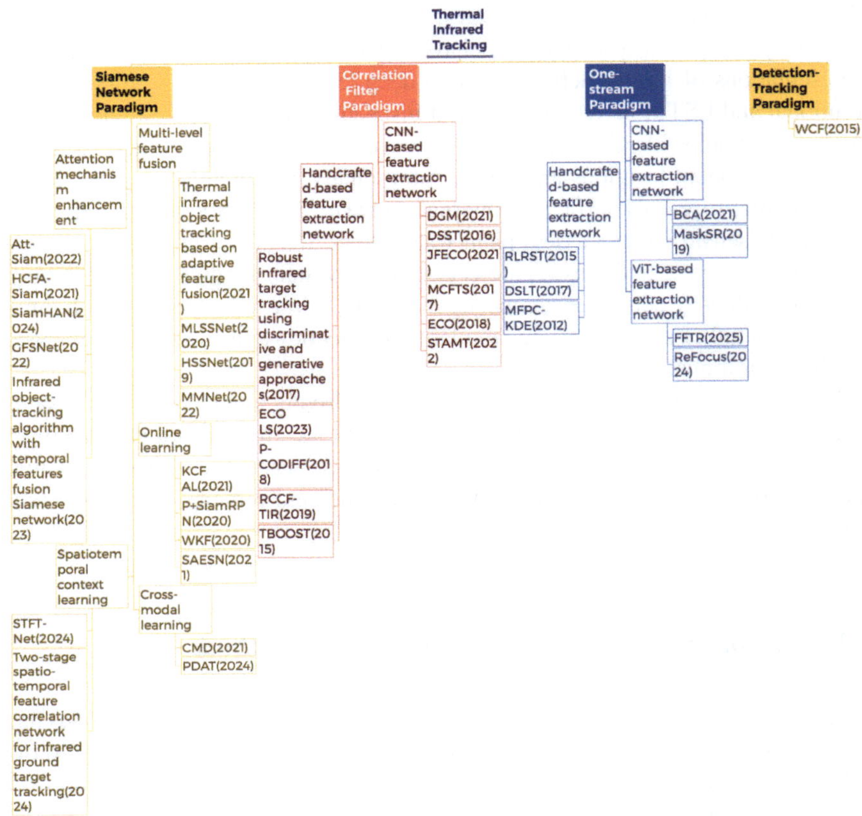

Fig. 3.7 Classification of VOT Method Based on Thermal Infrared

I Multi-Level Feature Fusion

Section 3.1.2.1 has already introduced multi-level feature fusion, so it will not be repeated here. Next, the focus will be on specific tracking methods directly.

The method in [199] builds on SiamRPN++ by introducing an adaptive feature fusion module. This module assigns channel weights to ResNet-50 features. It combines semantic information with fine details. It also uses the CIoU loss function. This improves the accuracy of bounding box regression.

MLSSNet [124] combines a structural similarity network for local texture matching and a semantic similarity network for global semantic matching. It uses relative entropy to adaptively integrate two types of similarity scores, boosting the ability to distinguish similar distractors.

HSSNet [108] fuses multi-level convolutional features and employs a spatially aware network to enhance feature distinctiveness. It performs end-to-end similarity learning for target matching.

MMNet [125] achieves robust thermal infrared tracking with a two-layer feature model. The first layer discriminates different categories via a multi-classification network. The second layer captures fine-grained differences within the same category through holistic and pixel-level modules. These features complement each other and are jointly optimized in a multitask matching framework.

II Attention Mechanism Enhancement

Section 3.1.2.1 has already introduced attention mechanism enhancement, so it will not be repeated here. Next, the focus will be on specific tracking methods directly.

Att-Siam [111] introduces cross-attention between the template and detection branches. This allows feature interaction and enhancement. It also applies spatial attention to highlight key regions. This helps focus on important areas in the image. An anchor-free regression network is used for localization. An improved cross-correlation formula further boosts accuracy for small infrared targets.

HCFA-Siam [216] combines shallow spatial and deep semantic features through a hierarchical convolutional feature aggregation network. It emphasizes target-related channels with a channel attention module and uses a negative template pool to enhance distinction from similar backgrounds.

SiamHAN [232] incorporates an improved CIoU loss function into the Siamese network to boost bounding box regression accuracy. It uses a GCNet attention layer to extract local features. An ECANet attention mechanism fuses multi-level features. This design improves robustness against occlusion. It also helps handle target deformation more effectively.

GFSNet [23] features adaptive subnetworks with classification and position-adaptive blocks. They dynamically enhance feature scores from depth-wise separable cross-correlation. This increases the model's sensitivity to input changes. It also improves generalization ability across different scenarios.

The method in [167] employs element-level attention to weight historical frame features in channel and spatial dimensions. It uses a discriminative update strategy to manage the feature pool. High-confidence tracking results are selected dynamically. These results are then used to update the pool. This enriches target feature diversity and improves visibility.

III Online Learning

Section 3.1.2.1 has already introduced online learning, so it will not be repeated here. Next, the focus will be on specific tracking methods directly.

KCF_AL [205] integrates a correlation filter layer into the Siamese network. It detects background or decoy interference using the peak-to-sidelobe ratio of the response map. This triggers network fine-tuning to suppress interference responses. As a result, it iteratively optimizes both features and tracking results.

P+SiamRPN [260] constructs a three-channel prediction model (original target, current appearance, surrounding environment). It generates updated samples for SiamRPN. Templates are dynamically adjusted using refined historical samples. This enhances adaptability to changes in pedestrian appearance.

WKF [92] uses weighted kernel filtering during inference to dynamically update template branch features. Combined with MobileNetV2 and multi-RPN branches, it enables real-time tracking.

SAESN [223] employs dilated convolutions to enlarge the receptive field. It stores historical tracking results in a dynamic template library. The optimal template is selected based on confidence scores. This reduces reliance on the initial template.

IV Spatiotemporal Context Learning

This kind of network abandons the limitations of traditional single-space feature matching and instead constructs a more complex dual-branch or multi-branch architecture. One branch deals with static features, usually using a template from the initial frame to capture the target's appearance and spatial information. The other branch focuses on dynamic feature extraction, introducing temporal information to capture the target's movement patterns and state changes. This design enables the network to sense the target's current state. It also captures the historical trajectory at the same time. As a result, it handles scale changes, appearance variations, and occlusion more effectively in complex scenes.

Moreover, such networks often employ sophisticated feature fusion modules. These modules integrate spatial and temporal features across different levels to enhance the features' discriminatory power and robustness. This approach enables the network to accurately locate targets. Additionally, when targets are disturbed or partially occluded, the network can maintain stable tracking by leveraging its spatiotemporal perception capabilities.

STFT-Net [154] proposes a spatiotemporal fusion transformer that processes triplets of static template, dynamic template, and search region to extract features. The encoder's multi-head self-attention captures spatiotemporal dependencies, and the decoder's cross-attention interacts with target queries to generate similarity maps. By integrating salient point features for dynamic template updates, this approach effectively addresses target deformation and occlusion.

The two-stage architecture in [106] first uses optical flow to extract spatiotemporal context, combining appearance and motion features to enhance discriminability. In the second stage, a unidirectional trajectory feature correlation network models historical trajectories to predict target locations. This is combined with search region expansion and a "secondary judgment" mechanism to handle long-term occlusions.

V Cross-Modal Learning

The core characteristic of this category is addressing the domain shift between RGB and TIR images. It aims to transfer the strong feature-representation ability of RGB modality to TIR modality through feature alignment and knowledge transfer. As TIR data annotation is difficult and large-scale labeled datasets are lacking, these methods often rely on unlabeled paired RGB-TIR data. They use cross-modal learning techniques to reduce the distribution gap between the two modalities. In this way, cross-modal learning methods can improve the target recognition and positioning abilities of TIR trackers without relying on large amounts of labeled TIR data. The research focuses on cross-modal feature alignment. It aims to design effective alignment mechanisms. It also explores new knowledge-transfer techniques. These efforts boost the robustness and accuracy of TIR trackers.

CMD [171] creates a dual-branch cross-modal distillation framework. It uses a pre-trained RGB teacher network to guide the TIR student network in learning target representations. It uses center cropping, random cropping, and detection-based sampling. These methods generate semantically aligned RGB-TIR image patches. This enables unsupervised knowledge transfer.

PDAT [104] presents a progressive domain adaptation framework. Initially, adversarial training is utilized to achieve global alignment of RGB-TIR feature distributions. Subsequently, sub-domain refinement is conducted based on clustering. It integrates large-scale unlabeled TIR data with SAM-generated pseudo-labels. This combination is used for training. As a result, it effectively addresses cross-domain feature discrepancies.

3.2.2.2 Correlation Filter Paradigm

Chapter 2 and Sect. 3.1.2.2 have introduced the correlation filter paradigm, so no further elaboration will be provided here. Subsequently, based on the design of the feature extraction network, this paradigm can be further categorized into two types: CNN-based and handcrafted-based.

I CNN-Based Feature Extraction Network

Section 3.1.2.2 has already introduced CNN-based feature extraction networks, so it will not be repeated here. Next, the focus will be on specific tracking methods directly.

DGM [88] enhances the ECO framework with a dynamic graph memory mechanism. It builds a multi-scale memory graph network across frames. Through spatiotemporal connections, it aggregates historical context and dynamically combines memory nodes with current-frame features. This improves infrared pedestrian localization while balancing computational efficiency with online node-set updates and dynamic feature selection.

DSST [63] is a classic correlation filter tracker. It enhances infrared target tracking by fusing multi-channel features like HOG, Gist, and depth infrared signatures. Using ridge regression, it creates multi-channel filters and performs efficient correlation in the frequency domain. Its key innovation lies in separate moving-average updates for the numerator and denominator, combined with a multi-scale estimation module. This approach maintains real-time performance while effectively handling target scale changes.

JFECO [206] combines feature learning with correlation filtering. A shallow embedding network with contrastive center loss learns discriminative target representations online. It uses a two-stage optimization process. The first stage applies contrastive learning to separate target from background. The second stage uses correlation filtering for efficient localization. This design shows strong adaptability in aircraft tracking.

MCFTS [123] creates a multi-cue tracking system with deep feature fusion. It uses pre-trained CNNs to extract multi-level features and builds weak trackers.

Responses are fused via KL divergence, with online model updates and scale estimation.

ECO [247] overcomes thermal infrared data scarcity using image translation. pix2pix converts labeled RGB videos into synthetic thermal infrared data while retaining annotations. A deep CNN trained on this synthetic data extracts thermal infrared features, which are then embedded into the ECO correlation filter tracking framework. Combined with motion features, this enables efficient tracking.

STAMT [233] presents a structured target-aware model. It builds on correlation filtering. A target-aware module is used to enhance target saliency. A dynamic structural weight model evaluates region credibility. These steps help handle complex interferences.

II Handcrafted-Based Feature Extraction Network

Section 3.1.2.2 has already introduced handcrafted-based feature extraction networks, so it will not be repeated here. Next, the focus will be on specific tracking methods directly.

The method in [5] enhances infrared target tracking robustness by fusing discriminative and generative techniques. Within a discriminative framework, it combines kernelized correlation filters with gradient and channel-encoded features for target localization. An AdaBoost classifier with pixel-intensity features is used for target segmentation, outputting candidate positions via dual parallel paths. In the generative part, optimal positioning is determined by maximizing multi-frame template normalized cross-correlation similarity. Lucas–Kanade homogeneous transformation estimates target scale changes.

ECO_LS [239] refines tracking in two stages: initial coarse localization with ECO, followed by fine-grained segmentation of the coarse region with level set methods. Historical information encoding accelerates level-set convergence. Its variant, ECOHG_LS, integrates HOG and grayscale features. This improves target representation. It also boosts tracking performance on multiple thermal infrared benchmarks.

P-CODIFF [38] presents a part-based robust tracking strategy. It divides the target template into key parts, constructs part-level features with co-difference descriptors, and matches parts using Euclidean distance. A dynamic update mechanism reinitializes parts deviating from the tracking region. The target state is determined by fusing part positions.

RCCF-TIR [227] improves the continuous correlation filtering framework. It uses an adaptive feature fusion mechanism. It also introduces the Average Peak-to-Correlation Energy metric. This metric dynamically evaluates feature reliability. Based on this, it adjusts the weight distribution of multiple features.

TBOOST [62] maintains a set of correlation filters to enhance model representation. Each filter corresponds to a specific target appearance pattern, with switches and updates based on appearance changes. A "survival-of-the-fittest" mechanism controls computational complexity by eliminating redundant filters based on usage frequency. The target position is determined by fusing the outputs of all filters through weighted integration.

3.2.2.3 One-Stream Paradigm

Chapter 2 and Sect. 3.1.2.3 have introduced the one-stream paradigm, so no further elaboration will be provided here. Subsequently, based on the design of the feature extraction network, this paradigm can be further categorized into three types: CNN-based, Handcrafted-based, and ViT-based.

I CNN-Based Feature Extraction Network

Section 3.1.2.2 has already introduced CNN-based feature extraction networks, so it will not be repeated here. Next, the focus will be on specific tracking methods directly.

BCA [198] effectively reduces brightness flickering in infrared videos. It first uses fixed-interval sampling to get scene brightness parameters. Then, it applies Kalman filtering for dynamic brightness estimation. This allows real-time brightness correction. During implementation, it adjusts only the brightness of background-region pixels. It also applies per-frame smoothing. This reduces brightness changes between adjacent frames. This keeps the target regions stable. It greatly improves visual coherence. It also provides a steady image foundation for tracking tasks.

MaskSR [102] innovatively combines deep features with sparse representation. It first extracts deep features using a pre-trained VGG-Net19. Then, it uses a channel selection layer to generate a binary mask template. This divides the target region into reliable and unreliable sub-regions. It uses a mask-weighted sparse representation model. It also applies the ADMM optimization algorithm. This approach focuses on reconstructing reliable regions accurately. At the same time, it handles multi-constraint reconstruction problems. Further integrated with a modified particle filter framework, it employs a kernel-based correlation filtering mechanism for particle screening. This maintains tracking accuracy while significantly improving computational efficiency.

II Handcrafted-Based Feature Extraction Network

Section 3.1.2.2 has already introduced handcrafted-based feature extraction networks, so it will not be repeated here. Next, the focus will be on specific tracking methods directly.

RLRST [74] presents a particle-filter-based tracker. It fuses robust low-rank sparse representation, models complex noise via maximum likelihood estimation, and adjusts noise weights dynamically. The core is formulating tracking as a joint optimization of low-rank matrices and sparse coefficients. It constructs a target representation model with candidate samples and selects optimal tracking results through weight calculations.

DSLT [230] uses an online structural learning framework. It speeds up dense sample processing using Fourier transforms. It then converts image window features into response maps through linear filtering. Finally, it locates the target by finding the peak position. Its innovation lies in a joint update mechanism for feature and loss maps, dynamically adjusting classifier parameters. This achieves effective suppression of target deformation and background interference at 215FPS, balancing real-time performance and tracking accuracy.

MFPC-KDE [126] proposes a multi-feature fusion strategy. It builds a pseudo-color feature space with intensity, Gabor, and weighted entropy features. By extending traditional single-feature KDE to a multi-dimensional feature domain, it leverages feature complementarity to enhance target saliency. This improves feature-space separability between targets and backgrounds and overcomes tracking drift caused by infrared image texture scarcity.

III ViT-Based Feature Extraction Network

Section 3.1.2.3 has already introduced ViT-based feature extraction networks, so it will not be repeated here. Next, the focus will be on specific tracking methods directly.

FFTR [116] uses a frequency-channel attention mechanism. Thermal infrared images are transformed into the frequency domain via discrete cosine transformation to extract texture details. It uses a dynamic template reconstruction strategy. This strategy rebuilds the previous template with the current search region. It also improves attention calculations. As a result, it makes better use of temporal information. The network is trained end to end, combining contrastive learning and masked reconstruction to optimize detail capture and adaptability. This improves thermal infrared target tracking performance.

ReFocus [95] uses a top-down attention-refocusing mechanism. Initially, a feed-forward network processes thermal infrared queries and search frames to produce initial results. Then, a query-guided feature selection module enhances task-related features using salient target information. The filtered features are fed back into the encoder. This is done through a feedback path across different levels. They are then integrated into the feed-forward process using adaptive projection. Finally, a second feed-forward integration strengthens the features, achieving attention optimization for thermal infrared tracking.

3.2.2.4 Detection-Tracking Paradigm

Section 3.1.2.5 has introduced the detection-tracking paradigm, so no further elaboration will be provided here. Next, the focus will be on specific tracking methods directly.

WCF [74] achieves infrared target tracking with a two-stage collaborative mechanism. In the detection stage, it uses a low-rank representation-based sequential detection algorithm. This algorithm analyzes continuous frame data to separate background and target regions. In the filtering stage, it constructs a multi-feature weighted function. This function integrates grayscale, spatial, and motion features. Target localization is refined using a weighted correlation filter to generate a target confidence map.

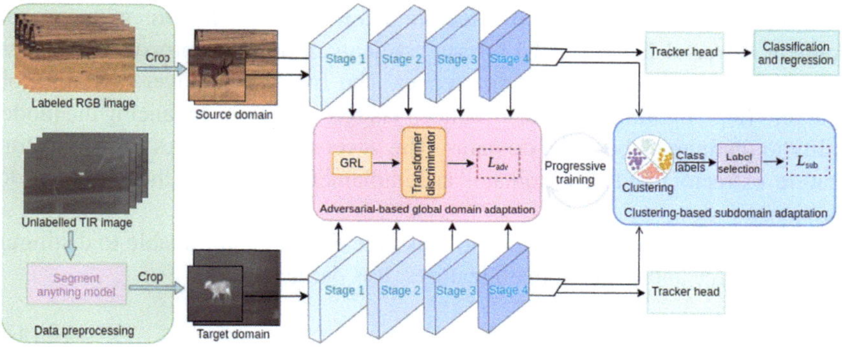

Fig. 3.8 The main architecture diagram of PDAT [104]

3.2.2.5 Typical Methods

I Siamese Network Paradigm Typical Method: PDAT [104]

Trackers trained on RGB datasets perform poorly on TIR data. This is mainly due to the lack of large-scale labeled TIR datasets. To solve this problem, the Progressive Domain Adaptation framework for TIR tracking (PDAT) is proposed. The framework is shown in Fig. 3.8. PDAT incorporates two key components: Adversarial-based Global Domain Adaptation (AGDA) and Clustering-based Sub-Domain Adaptation (CSDA). Progressive training is used to align domains. AGDA and CSDA gradually reduce the gap between TIR and RGB domains. They also learn fine-grained features that are invariant across domains.

Preliminary

The research team first collected a large-scale unlabeled TIR dataset with 2,549 image sequences and over 1.48 million frames, covering diverse scenarios and platforms. To tackle the lack of labels, they used the Segment Anything Model (SAM) to generate pseudo-labels. They optimized several processes to boost efficiency. For example, they segmented every tenth frame and raised the confidence threshold. These changes reduced preprocessing time from six days to two. As a result, they generated over nine million TIR training sample pairs. This is six times larger than the original dataset. This offline process provided high-quality, diverse inputs for subsequent modules, enhancing the model's feature transfer ability in complex scenarios.

AGDA Implementation Details

AGDA aligns the global distributions of RGB (source domain) and TIR (target domain) at the feature level via adversarial learning. Its core is a Transformer-based style discriminator, structured as follows:

$$D = \text{Linear}(\text{Transformer}(\text{GRL}(\mathbf{X}))) \tag{3.17}$$

Here, $\mathbf{X} \in \mathbb{R}^{H \times W \times C}$ is the feature map from the feature extractor. The Gradient Reversal Layer (GRL) inverts gradients during backpropagation, forcing the generator to confuse domain differences. The Transformer layer captures global style info (e.g., lighting, noise distribution) via self-attention, enhancing the model's ability to distinguish between domains. The adversarial objectives of the generator (feature extractor) and discriminator are implemented through the following loss functions:

The generator loss $\mathcal{L}_{adv\_G}$ aims to make the discriminator unable to distinguish between source and target domain features. It is defined as:

$$\mathcal{L}_{adv\_G} = (D(\mathbf{X}_t) - l_s)^2 + (D(\mathbf{Z}_t) - l_s)^2 \tag{3.18}$$

The discriminator loss $\mathcal{L}_{adv\_D}$ requires the discriminator to correctly classify the domain origin of features:

$$\mathcal{L}_{adv\_D} = \sum_{d=s,t} (D(\mathbf{X}_d) - l_d)^2 + (D(\mathbf{Z}_d) - l_d)^2 \tag{3.19}$$

During training, the generator and discriminator are optimized alternately. The generator minimizes $\mathcal{L}_{adv\_G}$ and the tracking task loss $\mathcal{L}_{track}$ (e.g., classification and regression losses) to adjust feature distributions. The discriminator minimizes $\mathcal{L}_{adv\_D}$ to improve domain classification. This adversarial process compels the generator to learn domain-invariant features, reducing the global distribution discrepancy between domains. Additionally, AGDA is applied across the four stages (Stages 1–4) of the ResNet50 backbone network, enhancing model adaptability through multi-level feature alignment.

CSDA Implementation Details

AGDA first achieves global feature alignment. Based on this, CSDA takes a further step. It generates pseudo-class labels through clustering. Then, it uses LMMD to align subdomain distributions between the RGB and TIR domains. This process tackles the issue of similar targets being hard to distinguish in tracking tasks. The key steps are as follows:

First, CSDA generates candidate pseudo-labels by calculating cross-domain feature correlations. For template features $\mathbf{Z}_d^m$ and search region features $\mathbf{X}_d^m$, a cross-correlation map is computed. These features come from stage m of the backbone network. Here, $d \in \{s, t\}$ denotes the domain.

$$\mathbf{F}_d^m = \mathbf{Z}_d^m \star \mathbf{X}_d^m \tag{3.20}$$

Then, $\mathbf{F}_d^m$ is clustered using K-means into C clusters to create pseudo-class labels. Given the varying clustering results across stages, CSDA uses a voting mechanism: it assigns linearly increasing weights (1–4) to the clustering results from stages 1 to 4 and selects the pseudo-label with the highest vote as the final class label.

Based on the pseudo-labels, CSDA employs LMMD to align the subdomain distributions. The subdomain alignment loss is defined as:

$$\mathcal{L}_{sub} = \frac{1}{C}\sum_{c=1}^{C}\Bigg[\sum_{i=1}^{n_s}\sum_{j=1}^{n_s} w_i^{sc} w_j^{sc} k(\mathbf{X}_i^s, \mathbf{X}_j^s) + \sum_{i=1}^{n_t}\sum_{j=1}^{n_t} w_i^{tc} w_j^{tc} k(\mathbf{X}_i^t, \mathbf{X}_j^t) \\ -2\sum_{i=1}^{n_s}\sum_{j=1}^{n_t} w_i^{sc} w_j^{tc} k(\mathbf{X}_i^s, \mathbf{X}_j^t)\Bigg] \tag{3.21}$$

This loss minimizes the distance between source and target domains in the same subclass within the domain while maximizing cross-domain differences. It forces the model to learn more discriminative fine-grained features.

Loss Function Implementation Details

The training is carried out in two stages. In the first stage, global domain adaptation and tracking are jointly optimized. The loss function combines adversarial loss and tracking task loss.

$$\mathcal{L}_{step1} = \mathcal{L}_{adv\_G} + \mathcal{L}_{adv\_D} + \mathcal{L}_{track} \tag{3.22}$$

Here, $\mathcal{L}_{adv\_G}$ and $\mathcal{L}_{adv\_D}$ from the AGDA module's adversarial learning: the generator confuses domain labels, while the discriminator classifies domain sources. $\mathcal{L}_{track}$ uses the baseline tracker's (e.g., SiamCAR) loss, including classification loss $\mathcal{L}_{cls}$, regression loss $\mathcal{L}_{reg}$, and centerness loss $\mathcal{L}_{cen}$, defined as:

$$\mathcal{L}_{track} = \lambda_1\mathcal{L}_{cls} + \lambda_2\mathcal{L}_{reg} + \lambda_3\mathcal{L}_{cen} \tag{3.23}$$

The weight coefficients λ_1, λ_2, and λ_3 are generally set to 1 to balance the different loss components.

In the second stage, subdomain adaptation is refined. The parameters of the global domain adaptation module are fixed. Only the subdomain alignment loss is optimized:

$$\mathcal{L}_{step2} = \mathcal{L}_{sub} \tag{3.24}$$

Here, $\mathcal{L}_{sub}$ from the CSDA module aligns cross-domain distributions of the same class via LMMD.

II Correlation Filter Paradigm Typical Method: STAMT [233]

As shown in Fig. 3.9, thermal infrared tracking faces two challenges. It is vulnerable to interference from similar objects. It also struggles with occlusion. To address

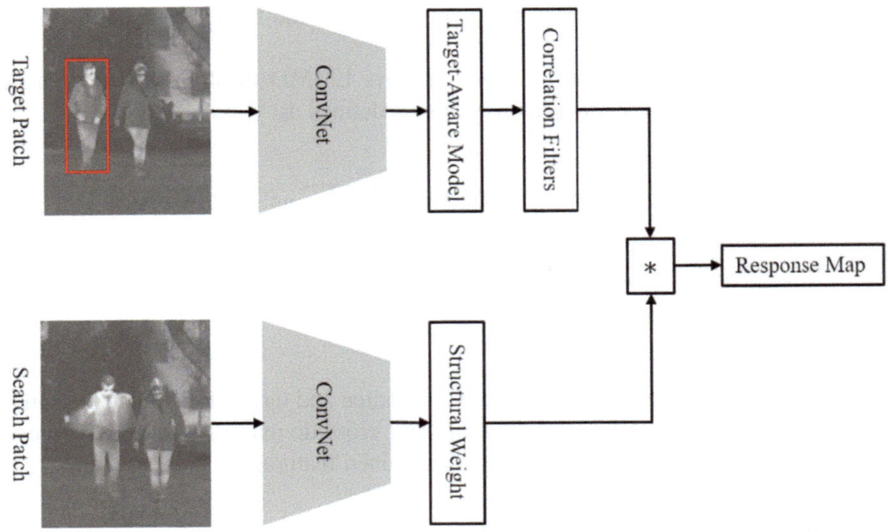

Fig. 3.9 The main architecture diagram of STAMT [233]

these issues, the Structural Target-Aware Model (STAMT) is proposed. The Target-Aware Model (BPTA) enhances target recognition by adjusting the correlation filter's training function, focusing more on the target. The Structural Weight Model (BPSW) improves tracking accuracy. It first divides the target search area into sub-regions. Then, it identifies reliable, unoccluded parts using response values. Finally, it increases the weights of these parts. Additionally, a base tracker is built as STAMT's underlying framework for feature extraction and initial response map generation:

Base Tracker Implementation Details
The base tracker's core components are a convolutional layer and a correlation filter layer. The convolutional layer, based on the conv4-3 layer of the VGG-16 network, extracts target features. The correlation filter layer generates a response map to locate the target. It optimizes an objective function to minimize the mean square error. The error is measured between training samples and Gaussian labels. A regularization term is added to prevent overfitting. This design draws on the CREST tracker but omits spatial and temporal residual modules for simplicity, laying the groundwork for subsequent integration of BPTA and BPSW.

BPTA Implementation Details
BPTA enables the model to focus more on the target region during correlation filter training. Specifically, this is achieved by adding an extra term related to the target training samples to the correlation filter's objective function, which is as follows:

$$w = \arg_w \min(\|w * x - y\|^2 - \eta \|w * \dot{x}\|^2 + \lambda \|w\|^2), \tag{3.25}$$

Here, $\|w * \dot{x}\|^2$ denotes the target-aware component. $\dot{x}$ is the target training sample, and η controls the significance of the target-aware model. This design strengthens the model's response to $\dot{x}$. It weakens responses to other samples. This helps distinguish the target from similar objects and the background. As a result, target recognition is enhanced.

Additionally, when integrating the target-aware model with a convolutional neural network, the objective function takes the form of:

$$L(w) = \mathcal{L}_w(x) - \eta\|\mathcal{F}_{\prime}(x)\|^2 + \lambda\|w\|^2, \tag{3.26}$$

Here, $\mathcal{F}_{\prime}(x) = w * \dot{x}$ represents the target-aware component. This allows the target-aware model to enhance feature expression in the target region. As a result, the tracker can accurately locate the target. It remains effective even in scenes with similar object interference.

BPSW Implementation Details

BPSW divides the target search area into sub-regions and determines whether each part is occluded by analyzing the response values in the response map. Specifically, it calculates the PSR (peak sidelobe ratio) for each sub-region to assess its reliability and compute the corresponding weight. The formula for PSR is as follows:

$$\text{PSR}_P = \frac{\max(R_P) - \mu_P}{\sigma_P}, \tag{3.27}$$

Here, R_P denotes the response output of the P -th sub-region, while μ_P and σ_P stand for the mean and standard deviation of the response values in that sub-region. Based on the PSR, the formula for calculating the weight is as follows:

$$\eta_P^t = \text{PSR}_P^t * \left|\max(R_P^t) - \max(R_P^{t-1})\right|, \tag{3.28}$$

Here, R_P^t and R_P^{t-1} denote the response outputs of the P-th sub-region in the current and previous frames, respectively. Ultimately, the structural weight model aggregates these weighted responses to generate the final response output, as formulated below:

$$R = \sum_{P=1}^{n} \eta_P * R_P, \tag{3.29}$$

Here, η_P denotes the weight of the Pth sub-region, R_P represents its response output, and n is the total number of sub-regions. By boosting the weights of unoccluded, reliable sub-regions, BPSW enhances tracking accuracy. This allows the tracker to precisely locate the target even when it is partially occluded.

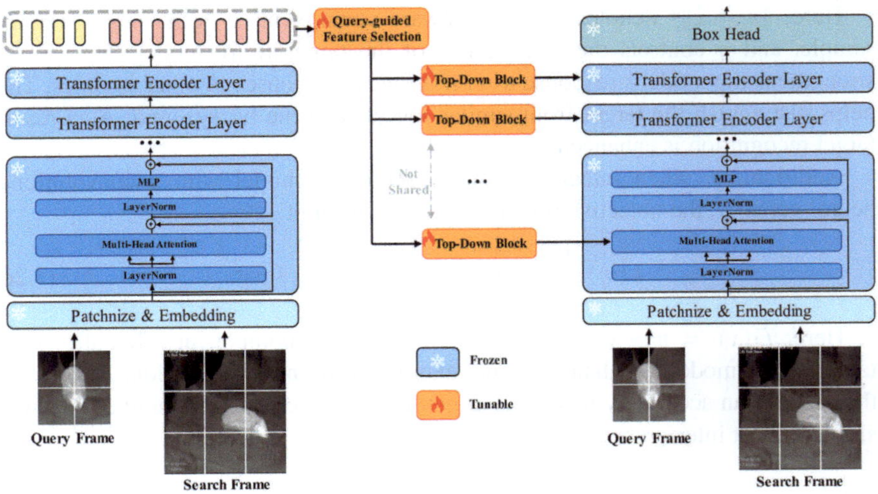

Fig. 3.10 The main architecture diagram of ReFocus [95]

III One-Stream Paradigm Typical Method: ReFocus [95]

As shown in Fig. 3.10, ReFocus is a novel parameter-efficient fine-tuning (PEFT) technique. It aims to solve the adaptation challenge in thermal infrared tracking. Deep tracking models are often pre-trained on large-scale RGB datasets. However, they perform poorly in thermal tracking due to limited training data. ReFocus addresses this issue effectively. Specifically, with all foundation model parameters frozen, only the query-guided feature selection and top-down attention modules are trained. This approach uses a top-down feedback mechanism. It injects task-specific features into the foundation model's attention mechanism. This improves the model's adaptability to thermal infrared tracking.

Foundation Tracker Implementation Details

The foundation tracker is based on OSTrack, which can jointly perform feature learning and relation modeling. It has a compact structure but delivers strong performance. OSTrack consists of a Transformer backbone network and a center-based bounding box prediction head network. The input includes the query image region and the search image region. These are split and flattened into image patches and projected into latent space via an embedding layer. After adding 1D position embeddings, the resulting token sequence goes through multiple Transformer encoder layers. Finally, the bounding box prediction head network uses the center and offset branches for object localization and the size branch for scale prediction.

Feature Selection Module Implementation Details

The query-guided feature selection module draws inspiration from the selective attention mechanism of the human visual system. It aims to modulate the foundation model's initial output features using prior information from the query frame (template frame). This enhances features relevant to the thermal infrared tracking task and suppresses irrelevant ones. The process involves two main steps: spatial feature selection and channel feature selection.

During spatial feature selection, average pooling is applied to the query token. This produces an average query token $k_{\text{avg}}^L \in \mathbb{R}^{1\times D}$. It represents the overall features of the query frame. Then, k_{avg}^L is element-wise multiplied with a learnable embedding $p_e \in \mathbb{R}^{1\times D}$. This operation generates a target-aware representation $r_s \in \mathbb{R}^{1\times D}$, as follows:

$$r_s = k_{\text{avg}}^L \circ p_e \tag{3.30}$$

Here, $\circ$ denotes element-wise multiplication. Next, the cosine similarity between the target-aware representation r_s and the token sequence K_{zx}^L is calculated. The resulting similarity scores serve as a weighted mask. This mask is used to weigh the original token sequence spatially. The similarity scores are first broadcast to the channel dimension. Then, a token-wise (spatial Hadamard) product is performed with the original token sequence. This produces the spatially weighted token sequence $\hat{K}_{zx}^L$.

$$\hat{K}_{zx}^L = \text{Mul}\left(r_s, K_{zx}^L\right) \otimes K_{zx}^L \tag{3.31}$$

In the channel feature selection phase, global max-pooling is applied to the query token. This yields the maximum value $k_{\max}^L \in \mathbb{R}^{1\times D}$. It represents the most prominent feature information of the query frame. Then, $k_{\max}^L$ is multiplied with a learnable weight matrix $p_w \in \mathbb{R}^{D\times D}$. This generates the channel importance representation $r_c \in \mathbb{R}^{1\times D}$. Finally, r_c is broadcast to the spatial dimension. It then performs a channel-wise (Hadamard) product with $\hat{K}_{zx}^L$. This produces the channel-selected token sequence $\tilde{K}_{zx}^L$:

$$\tilde{K}_{zx}^L = r_c \odot \hat{K}_{zx}^L \tag{3.32}$$

Top-Down Blocks Implementation Details

The feature sequence $\tilde{K}_{zx}^L$ from the query-guided feature selection module is transformed via "top-down blocks" with a bottleneck structure. These blocks, containing two linear layers, compress the feature dimension from D to a hidden dimension C and then restore it to D. The transformed features are then injected into each layer of the Transformer encoder.

In the ith layer of the Transformer encoder, the input features are composed of two parts: the output features K^{i-1} from the $i-1$th layer and the features $\tilde{K}_{zx}^L$ transformed by the top-down block ψ_{td}^{i-1}. To incorporate top-down signals into multi-head attention (MHA), ReFocus modifies the value (V) input. It adds the top-down signals to V. The query (Q) and key (K) remain unchanged. This allows each encoder layer to enhance its focus on task-related features using high-level information.

For the ith layer of the Transformer encoder, the input feature K^i is:

$$K^i = T^{i-1}\left(K^{i-1}, \psi_{td}^{i-1}\left(\tilde{K}_{zx}^L\right)\right) \tag{3.33}$$

Here, T^{i-1} is the $i-1$th Transformer encoder layer, and ψ_{td}^{i-1} is the corresponding top-down block.

In the multi-head attention mechanism, the query, key, and value are calculated as:

$$Q, K, V = W_Q x^{bp}, W_K x^{bp}, W_V(x^{bp} + x^{td}) \tag{3.34}$$

Here, x^{bp} is the bottom-up input, x^{td} is the top-down signal from $\psi_{td}^{i-1}(\tilde{K}_{zx}^L)$, and W_Q, W_K, W_V are the weight matrices.

3.2.3 Thermal Infrared VOT Methods Performance Comparison

Table 3.2 summarizes the performance of thermal-infrared-based object trackers from 2019 to 2024. It covers results across different benchmarks. The listed methods include Siamese networks, correlation filters, and one-stream paradigms. The table uses metrics like EAO and success rates (A and R) for evaluation. It includes benchmarks such as VOT-TIR 2015–2017, LSOTB-TIR, and PTB-TIR. Each method's strengths and weaknesses are highlighted. This helps researchers improve thermal-infrared tracking techniques. For instance, Siamese-network-based STFT-Net shows remarkable EAO in recent benchmarks. Some correlation filter and one-stream network methods also perform well in early or specific benchmarks, though some have lower metric values. Overall, the table reflects the development trend in thermal-infrared VOT and the performance characteristics of different methods in various scenarios.

3.2.4 Summary and Challenges

This section concentrates on the pivotal domain of thermal infrared object tracking. It encompasses the imaging mechanism, distinctive benefits, and broad applications of the thermal infrared modality. Delving deeper, it dissects the four primary paradigms of thermal infrared VOT: Siamese networks, correlation filters, one-stream networks, and detection-tracking frameworks. For each paradigm, the section highlights key methods and research focuses. In Siamese networks, it covers multi-

Table 3.2 Comparison of the performance of different Thermal-based trackers from 2019 to 2024

			VOT-TIR 2015			VOT-TIR 2016			VOT-TIR 2017			LSOTB-TIR		PTB-TIR	
Paradigms	Trackers	Publication	EAO	A	R	EAO	A	R	EAO	A	R	SR	PR	SR	PR
Siamese network	STFT-Net[154]	KBS2024	–	–	–	–	–	–	–	–	–	0.704	0.821	–	–
	HSSNet[108]	KBS2019	0.311	0.67	0.23	0.268	0.56	0.629	0.26	0.58	3.34	0.436	0.566	0.468	0.689
	MLSSNet[124]	TMM2020	0.327	0.57	0.25	0.294	0.564	0.667	0.278	0.56	3.13	0.457	0.593	0.516	0.734
	HCFA-Siam[216]	TIM2021	–	0.653	–	–	–	–	–	–	–	–	–	–	–
	GFSNet[23]	IP&T2022	0.365	0.66	0.22	–	–	–	–	–	–	0.685	0.797	–	–
Correlation filter	MCFTS[123]	KBS2017	0.218	0.59	0.41	–	–	–	–	–	–	0.475	0.633	0.498	0.68
	DGM [88]	SPL 2021	–	–	–	–	–	–	–	–	–	–	–	0.639	0.846
	ECO LS[239]	IVP2022	0.333	0.664	0.836	0.323	0.645	0.772	–	–	–	–	–	0.647	0.867
	STAMT[233]	NC 2022	–	–	–	–	–	–	–	–	–	0.577	0.713	0.574	0.781
One stream	MMNet[125]	AAAI2020	0.344	0.62	0.21	0.307	0.61	0.677	0.33	0.58	2.89	–	–	0.558	0.785
Other	MaskSR[102]	RS 2019	–	–	–	0.26	0.515	0.588	–	–	–	–	–	–	–

level feature fusion and attention mechanism enhancement. In correlation filters, it introduces innovative CNN-based and handcrafted feature extraction. Moreover, it spotlights several typical thermal infrared tracking methods. These methods show unique strengths in addressing key challenges. They help overcome issues like data annotation limitations, feature extraction, and tracking accuracy. These contributions lay a robust groundwork for the future advancement and practical implementation of thermal infrared tracking technologies.

Nonetheless, thermal infrared VOT applications are still confronted with the following challenges:

1. **Data annotation and volume insufficiency**: Large-scale labeled thermal infrared datasets are scarce, restricting the application of supervised learning methods and the training effectiveness of models. Although some methods expand data via SAM-generated pseudo-labels, issues like annotation inaccuracy still exist.
2. **Difficulty in feature representation and transfer**: There are significant differences between thermal infrared and RGB image features. Directly transferring feature representation methods from RGB modalities yields poor results. It is necessary to design effective cross-modal feature alignment mechanisms to narrow the domain distribution gap.
3. **Challenge of improving tracking robustness**: Complex scenarios (such as target occlusion, deformation, and scale changes) affect tracking stability. The robustness of models against these situations needs to be enhanced. Some methods, despite having structural target perception models, may still fail in the face of severe occlusion and rapid deformation.
4. **Dilemma of balancing real-time performance and efficiency**: Improving the operational speed of algorithms to meet real-time application requirements while ensuring tracking accuracy is challenging. Some methods, although parameter-efficient, still need to optimize real-time performance for higher resolutions or more complex scenarios.
5. **Complexity of multi-modal fusion**: In cross-modal learning, effective fusion of thermal infrared and other modalities is a key challenge. Combining thermal infrared with RGB can enhance tracking performance. How to fully use their advantages remains an urgent problem. The adaptability and stability of current fusion strategies across different scenarios still need further improvement.

3.3 LiDAR-based VOT

3.3.1 Modal Characteristics and Application Scenarios

With the continuous advancement of 3D acquisition technology, the application of 3D sensors such as LiDAR and RGB-D cameras has become increasingly widespread [69]. Point cloud data, as powerful spatial representation modality, have gradually gained attention. Compared to traditional 2D images, point clouds inherently possess advantages such as resistance to perspective distortion and occlusion. They directly encode the geometric structure of the environment through a set of unordered 3D points. Each point typically contains spatial coordinates (x, y, z) and may also include additional features such as intensity or color.

Point cloud data are unstructured and sparse. This makes it hard for traditional deep learning models that are designed for regular grids. To address this issue, researchers have developed various innovative learning paradigms, such as PointNet [151], PointNet++ [152], and KPConv [179], enabling neural networks to process raw point sets directly without relying on voxelization or meshing. This paradigm shift has significantly advanced 3D perception tasks, including object detection, semantic segmentation, and object tracking. It has demonstrated great potential for practical applications.

Key advantages of leveraging point cloud data include:

1. **High-precision 3D spatial representation**: Point cloud data provide a high-precision three-dimensional spatial description of objects and environments. The 3D coordinates (x, y, z) contained in each point can accurately represent the geometric shape of an object's surface. This capability is valuable for generating accurate 3D models in applications like autonomous driving, robotics, and environmental modeling [162].
2. **Rich geometric information**: The advantages of point cloud data are significant. They not only capture depth information but also provide detailed geometric insights through spatial distribution and shape [151]. This makes them crucial for tasks like object recognition, segmentation, and modeling. Compared to traditional 2D images, point clouds offer a more direct representation of an object's 3D form.
3. **Robustness in Complex Environments**: Point cloud data show greater robustness in challenging conditions [266] like low light, fog, and rain compared to 2D images. For example, LiDAR point clouds provide accurate depth information even at night or in poor visibility, such as fog, rain, or snow, where traditional cameras struggle. Unlike camera-based sensors, LiDAR does not depend on lighting and works effectively in both day and night, as well as in adverse weather [9]. It reliably captures high-quality depth data in dynamic environments, including haze, rain, or snow.
4. **Invariance of Point Clouds to Rotation and Translation**: Point cloud data demonstrate inherent invariance to rotation and translation [152]. This characteristic is due to the fact that point clouds represent object surfaces using 3D coordinates (x, y, z), independent of any specific viewpoint or fixed coordinate system. Therefore, regardless of how an object rotates or translates, the point cloud representation remains consistent, providing reliable support for tasks such as object recognition, tracking, and 3D reconstruction in dynamic environments.

In light of the unique advantages of point cloud data mentioned above, it has demonstrated tremendous application potential across various fields. The high-precision 3D spatial representation, rich geometric information, robustness in complex environments, and invariance to rotation and translation make point clouds an ideal choice for handling dynamic scenes. These characteristics not only enhance the accuracy of tasks such as object detection, semantic segmentation, and object tracking but also provide a solid foundation for practical applications:

- Object Tracking in Autonomous Driving: In autonomous driving, point cloud data typically originate from LiDAR sensors and are used to detect and track various targets on the road, such as pedestrians, vehicles, and obstacles. LiDAR point clouds excel in challenging environments, including low-light conditions, rain, snow, and fog, providing high-precision depth information. This capability effectively addresses the limitations of traditional vision-based sensors in complex scenarios.
- Object Tracking in Robotics and Mobile Devices: In the realm of robotics navigation and path planning, point cloud data play a crucial role in real-time tracking of target objects within the environment. Utilizing sensors such as LiDAR, robots are capable of acquiring precise positional information of targets in dynamic settings. This capability enables them to navigate and avoid obstacles based on the gathered data.
- Augmented Reality and Virtual Reality: In augmented reality (AR) and virtual reality (VR) applications, point cloud data can be used for real-time object tracking and environment modeling. By leveraging point cloud data, systems are able to efficiently construct and update virtual environments, thereby providing a more immersive experience.

3.3.2 Inductive Classification and Typical Methods

Since the advent of the pioneering 3D LiDAR-based tracker SC3D [56], 3D VOT based on LiDAR has become a rapidly developing research area in recent years. All related models are designed using deep learning methods. As shown in Fig. 3.11, these methods can be broadly categorized into three types: Siamese-based methods, motion-modeling methods, and one-stream Transformers methods. Next, the current approaches are introduced from these three perspectives.

I Siamese-Based Methods

Refer to Chap. 2, Siamese-based methods extract point cloud features from template and search regions using a shared backbone network and perform target matching through similarity learning. Most existing tracking methods fall into this category. This approach is inspired by traditional RGB-based object tracking methods and usually creates seed points for locating targets by matching features.

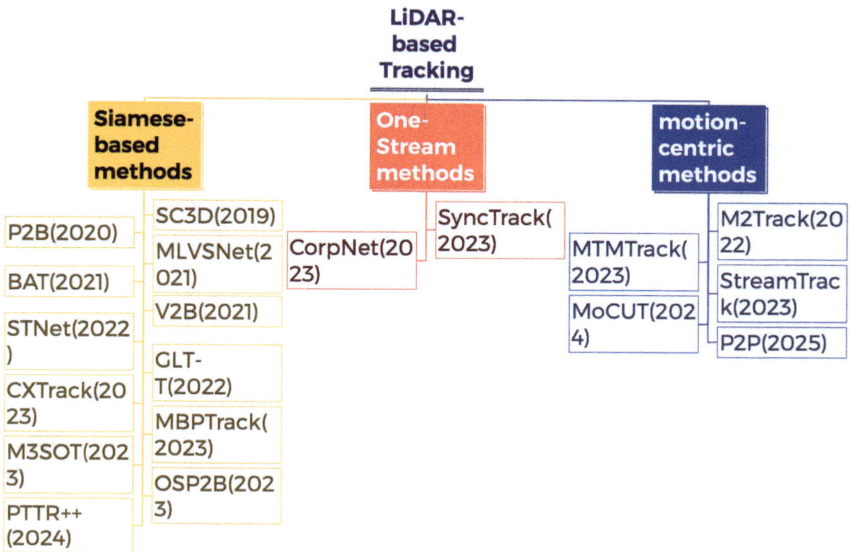

Fig. 3.11 Develop lineage of LiDAR-based trackers

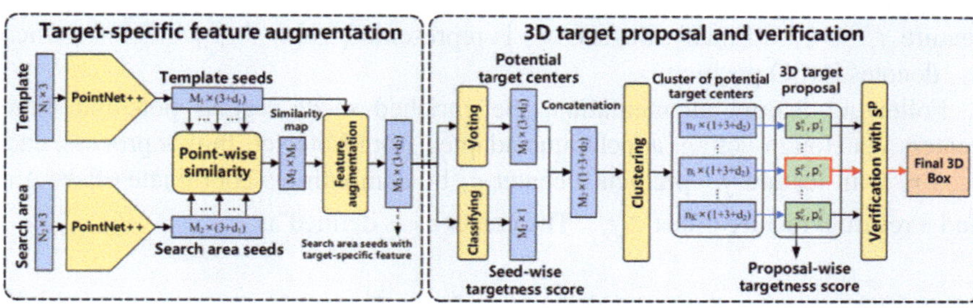

Fig. 3.12 The main architecture diagram of P2B [153]

As a pioneer in the field of 3D VOT in point clouds, SC3D [56] introduces the 3D Siamese tracker in an innovative way. It achieves precise tracking by extensively comparing the target template with the numerous candidate blocks generated by the current frame.

However, this approach requires multiple forward passes by the model, resulting in significant computational overhead. To address this issue, as shown in Fig. 3.12, P2B [153] introduces a more efficient approach by leveraging a Region Proposal Network (RPN) for proposal generation and utilizing Hough voting to produce tracking predictions.

Unlike traditional methods that rely on RGB-D data or exhaustive 3D searches, the proposed P2B network locates potential target centers in a 3D search area with target information. It then jointly performs point-driven 3D target proposal and verification, avoiding the need for a computationally expensive exhaustive search.

The method begins by sampling seeds from point clouds in both the template and search area, leveraging a modified PointNet++ backbone to process these inputs. The search area seeds are enriched through a permutation-invariant target-specific feature augmentation process. This process integrates target clues from the template. It includes 3D position coordinates to preserve spatial geometry, point-wise similarity with template seeds to uncover local tracking patterns, and the encoded global feature of the target. This augmentation is mathematically formalized by computing the point-wise similarity using cosine distance, expressed as

$$\mathrm{Sim}_{j,i} = \frac{f_{q_i}^{\mathrm{T}} \cdot f_{r_j}}{\left\| f_{q_i} \right\|_2 \cdot \left\| f_{r_j} \right\|_2}, \quad \forall q_i \in Q, r_j \in R, \tag{3.35}$$

where $\mathrm{Sim}_{j,i}$ represents the similarity between a template seed q_i and a search area seed r_j, with f_{q_i} and f_{r_j} denoting their respective feature vectors. The normalization ensures scale invariance.

To address the instability caused by the disorder of template seeds Q, a symmetric function such as Maxpool is applied. This function augments $\mathrm{Sim}_{j,i}$ with spatial coordinates and features from Q. The augmented data form a tensor, which is then processed by a network Φ (MLP-Maxpool-MLP) to generate the target-specific feature $f_{r_j}^{\mathrm{t}} \in \mathbb{R}^{d_2}$. Finally, the seed r_j is represented as $[x_{r_j}; f_{r_j}^{\mathrm{t}}] \in \mathbb{R}^{3+d_2}$, where x_{r_j} denotes its 3D position.

Following feature augmentation, the enriched seeds regress potential target centers via Hough voting, a technique adapted from VoteNet. In this process, each seed r_j with feature $f_{r_j}^{\mathrm{t}}$ predicts a center c_j by computing a coordinate offset Δx_j and a residual feature offset $\Delta f_{r_j}^{\mathrm{t}}$. The center c_j is defined as

$$x_{c_j} = x_{r_j} + \Delta x_{r_j}, \quad f_{c_j} = f_{r_j}^{\mathrm{t}} + \Delta f_{r_j}^{\mathrm{t}}, \tag{3.36}$$

with the regression loss given by

$$L_{\mathrm{reg}} = \frac{1}{M_{\mathrm{tc}}} \sum_j \left\| \Delta x_j - \Delta gt_j \right\| \cdot \mathrm{I}\left[r_j \text{ on target}\right], \tag{3.37}$$

where Δgt_j is the ground-truth offset, $\mathrm{I}(\cdot)$ indicates training only on seeds on the target surface. And M_{tc} is the number of trained seeds. Each potential target center is further enhanced with a seed-wise targetness score s_j^s. It is derived from a MLP and trained with a binary cross-entropy loss L_{cla}, which regularizes feature learning and strengthens center representation. The updated center representation becomes $[s_j^s; x_{c_j}; f_{c_j}] \in \mathbb{R}^{1+3+d_2}$.

For target proposal, each center c_j clusters neighboring seeds within a radius R using ball query to form T_j^{t}. To improve efficiency, a subset of K centroids is sampled. These clusters are then processed by another network Θ (MLP-Maxpool-

MLP), which generates K target proposals p_j^{t} along with proposal-wise targetness scores s_j^{p}, expressed as

$$\left\{p_j^{\mathrm{t}}, s_j^{\mathrm{p}}\right\} = \Theta\left(T_j^{\mathrm{t}}\right), \tag{3.38}$$

where p_j^{t} includes offsets for 3D position and rotation.

Following P2B, numerous methods have been proposed that build upon its framework to further advance 3D VOT in point clouds. For instance, BAT [259] proposes a box-aware feature fusion module to strengthen target-specific feature representation. Meanwhile, MLVSNet [201] adopts the CBAM [203] module to enhance vote cluster features by incorporating channel-wise and spatial attention mechanisms. V2B [86] performs Voxel-to-BEV transformation for object localization on the densified feature map. Subsequently, STNet [85] constructs a pure Transformer-based backbone to replace PointNet++ [85] for feature extraction and proposes an independent Transformer-based matching network, which consists of self-attention and cross-attention modules, to enhance the decoded features. GLT-T [144] proposes a novel Global-Local Transformer voting algorithm for 3D object tracking on point clouds, which enhances the quality of 3D proposals and improves the model's focus on crucial seed points through a cascaded global-local Transformer module and a simple yet effective training strategy, thereby boosting tracking performance. OSP2B [145] introduces a one-stage point-to-box network to improve the 3D Siamese paradigm. CXTrack [212] exploits contextual information across consecutive frames to improve tracking accuracy. Compared to CXTrack, MBPTrack [213] introduces a memory mechanism to leverage spatial and temporal context information from historical frames, propagating target cues to the current frame through a Transformer-based module. Additionally, MBPTrack proposes BPLocNet, a network that uses bounding box priors for coarse-to-fine localization to handle targets of varying sizes. These improvements enable MBPTrack to better address challenges such as target appearance changes and occlusions, thereby enhancing tracking performance.

As shown in Fig. 3.13, M3SOT [120] proposes a comprehensive framework incorporating multi-frame, multi-field, and multi-space components. Specifically, it is designed for 3D VOT tasks. The framework thoroughly analyzes the necessity of each module and reveals how to construct tasks to effectively address the VOT problem.

M3SOT starts by redefining the tracking problem to predict the target's relative motion $(\Delta x, \Delta y, \Delta z, \Delta\theta)$ using multiple historical frames $\boldsymbol{P}^{t-K:t-1}$ and the current frame $\boldsymbol{P}^t$. In this formulation, $\boldsymbol{M}^i \in \mathbb{R}^{N\times 1}$ represents the targetness mask that identifies tracked points. It is initialized to 0.5 for $\boldsymbol{P}^t$, as the target states are initially unknown. The framework adopts a many-to-one matching scheme. It concatenates point features $\boldsymbol{F}^i \in \mathbb{R}^{N\times C}$ and masks $\boldsymbol{M}^i$ from both the template set $\boldsymbol{P}^{t-K:t-1}$ and the search area $\boldsymbol{P}^t$. A shared DGCNN-based backbone is used to extract local geometric features. This is done through range sampling and EdgeConv layers.

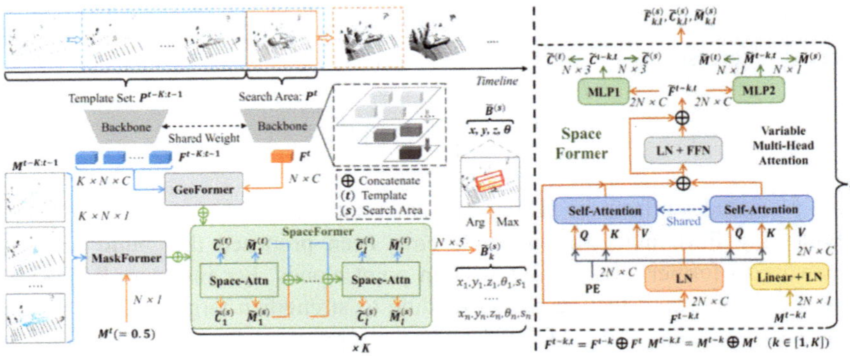

Fig. 3.13 The main architecture diagram of M3SOT [120]

Specifically, the process is expressed as $\boldsymbol{P}_s = RS(\boldsymbol{P}_0)$ and $\boldsymbol{F}_s = DGCN(\boldsymbol{F}_0)$. Here, RS preserves point relationships without additional computation, while $DGCN$ aggregates features across multiple receptive fields to address point cloud sparsity.

The core of M3SOT lies in its hybrid transformer architecture, comprising GeoFormer, MaskFormer, and SpaceFormer modules. GeoFormer processes geometric features $\boldsymbol{F}^{t-K:t} = \boldsymbol{F}^{t-K} \oplus \boldsymbol{F}^t$ to represent the template set and search area holistically. MaskFormer encodes point-box relationships using a mask $m_i^{(t)} = \begin{cases} 0, & p_i^{(t)} \text{ not in } \boldsymbol{B}^{(t)}, \\ 1, & p_i^{(t)} \text{ in } \boldsymbol{B}^{(t)}, \end{cases}$ similar to positional encoding, enhancing target-specific cues. SpaceFormer integrates these inputs via a variable multi-head attention mechanism, normalizing features as $\overline{\boldsymbol{F}}^{t-K:t} = LN(\boldsymbol{F}^{t-K:t})$, and computing attention with query $\boldsymbol{Q} = \boldsymbol{K} = \overline{\boldsymbol{F}}^{t-K:t} + PE$, key $\boldsymbol{K}$, and value $\boldsymbol{V} = \overline{\boldsymbol{F}}^{t-K:t}$. The attention is formulated as

$$\tilde{\boldsymbol{F}}^{t-K:t} = \boldsymbol{F}^{t-K:t} + MHA(\boldsymbol{Q}, \boldsymbol{K}, \boldsymbol{V}) + MHA(\boldsymbol{Q}, \boldsymbol{K}, \boldsymbol{ME}), \tag{3.39}$$

where MHA denotes multi-head attention, and single-head attention for subspace i is $\text{Attn}(\boldsymbol{Q}_i, \boldsymbol{K}_i, \boldsymbol{V}_i) \equiv \text{softmax}\left(\frac{\boldsymbol{Q}_i \boldsymbol{K}_i^T}{\sqrt{d_h}}\right) \boldsymbol{V}_i$ with $d_h = C/H$. The network depth L and heads H vary with template inputs, refining features layer-wise as $\tilde{\boldsymbol{F}}_i^{t-K:t} = \tilde{\boldsymbol{F}}_{i-1}^{t-K:t} + FFN(LN(\tilde{\boldsymbol{F}}_{i-1}^{t-K:t}))$, where $FFN(x) = \text{ReLU}(xW_1 + b_1)W_2 + b_2$. Intermediate tasks predict masks $\tilde{\boldsymbol{M}}_{k,i}^{(t,s)} = MLP_m(\tilde{\boldsymbol{F}}_i^{t-K:t})$ and centers $\tilde{\boldsymbol{C}}_{k,i}^{(t,s)} = MLP_c(\tilde{\boldsymbol{F}}_i^{t-K:t})$, with the final layer feeding $\tilde{\boldsymbol{F}}_{k,i}^{(s)}$, $\tilde{\boldsymbol{C}}_{k,i}^{(s)}$, and $\tilde{\boldsymbol{M}}_{k,i}^{(s)}$ to X-RPN for bounding box prediction $\tilde{\boldsymbol{B}}_k^{(s)} = X-RPN(\tilde{\boldsymbol{F}}_{k,i}^{(s)}, \tilde{\boldsymbol{C}}_{k,i}^{(s)}, \tilde{\boldsymbol{M}}_{k,i}^{(s)})$, selecting the maximum confidence score across K templates.

This design enables M3SOT to propagate target cues efficiently, leveraging historical context and multi-scale features to handle occlusions and sparsity.

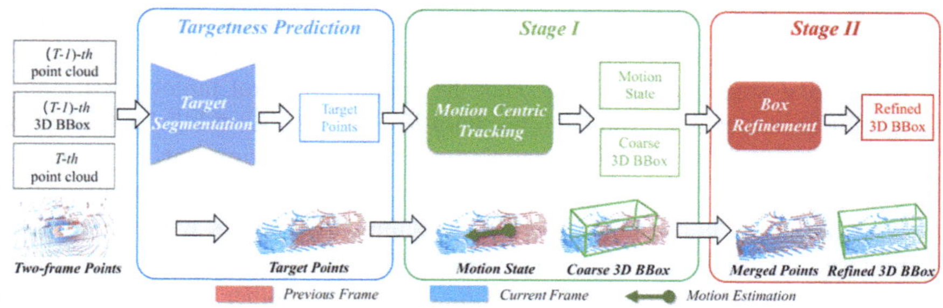

Fig. 3.14 The main architecture diagram of M2Track [258]

II Motion-Modeling Methods

Motion-modeling methods focus on leveraging the target's state in the previous frame to infer relative motion offsets in the current frame. These methods model the target's motion state to predict its position changes in the current frame and track it based on historical motion information. This approach improves the accuracy and robustness of target tracking in dynamic scenes.

As shown in Fig. 3.14, M2Track [258] introduced the motion-centric paradigm that located the target by estimating the relative motion between two frames and achieved remarkable performance improvement. Given an input LiDAR sequence and the 3D bounding box of the target in the first frame, M2Track aims to localize the target frame by frame through explicit motion modeling. At timestamp t ($t > 1$), assuming the target bounding box B_{t-1} at frame $t-1$ is known, with two consecutive frames P_t and P_{t-1}, as well as the target bounding box B_{t-1} in P_{t-1}, M2Track predicts the relative target motion (RTM) between these successive frames. Considering the alignment of targets with the ground plane, only 4 degrees of freedom (DOF) are considered for RTM instead of 6DOF. Specifically, a 4DOF RTM $M_{t-1,t} \in R^4$ defined between two target bounding boxes in frame t and $t-1$ includes translation offsets $(\Delta x, \Delta y, \Delta z)$ and yaw offset $\Delta\theta$. This process can be formulated as follows:

$$F : R^{N_t \times C} \times R^{N_{t-1} \times C} \times R^7 \rightarrow R^4, F(P_t, P_{t-1}, B_{t-1}) \rightarrow (\Delta x, \Delta y, \Delta z, \Delta\theta); \tag{3.40}$$

with the predicted RTM $M_{t-1,t}$, the target bounding box in P_t can be obtained using rigid body transformation:

$$B_t = Transform(B_{t-1}, M_{t-1,t}). \tag{3.41}$$

To achieve this, M2Track initially segments target points from their surroundings by exploiting spatial–temporal relations between two frames. Taking two consecutive frames P_t and P_{t-1} along with the target bounding box B_{t-1} as inputs, M2Track constructs a spatial–temporal point cloud $P_{t-1,t} \in R^{(N_{t-1}+N_t)\times 4}$ by adding a temporal channel to each point and merging them together. A prior-targetness map $S_{t-1,t} \in R^{N_{t-1}+N_t}$ is created to indicate the target's location in $P_{t-1,t}$.

At the first stage, M2Track encodes the spatial–temporal target point clouds $P_{t-1,t}$ into an embedding using another PointNet encoder. A multilayer perceptron (MLP) is applied on top of the embedding to obtain the motion state of the target, which includes a 4D RTM $M_{t-1,t}$ and 2D binary classification logits indicating whether the target is dynamic. To reduce accumulation errors while performing frame-by-frame tracking, a refined previous target bounding box B_{t-1} is generated by predicting its RTM with respect to B_{t-1} through another MLP. If the target is classified as dynamic, the current target bounding box B_t is calculated using Eq. 3.41.

Subsequently, at the second stage, M2Track refines the bounding box using a motion-assisted shape completion strategy. By aggregating targets from two successive frames based on the predicted motion state, a denser target point cloud is formed. The transformed point cloud is merged with the current frame's point cloud to form a denser point cloud $\hat{P_t} \in R^{(M_{t-1}+M_t)\times 3}$. After transforming $\hat{P_t}$ from the world coordinate system to the canonical coordinate system defined by B_t, a PointNet is applied to regress another RTM with respect to B_t. Finally, the refined target bounding box B_t is obtained by applying Eq. 3.40 on the regressed RTM and B_t.

MTM-Tracker [113] enhances motion modeling while incorporating appearance matching. It first uses historical bounding boxes as motion cues to roughly locate the target. Then, it employs a feature interaction module to extract motion-sensitive features from consecutive point clouds and aligns these features to strengthen the representation of the target trajectory. StreamTrack [131] focuses on capturing the long-range continuous motion attributes of the tracked object by constructing a repository of multiple historical features to interact with the current frame. MoCUT [147] adopts a novel strategy by using a single model for training and testing across all categories, which differs from previous methods that relied on category-specific models. This approach is based on a motion-centric 3D VOT paradigm. It achieves target tracking by integrating point cloud data from historical frames and extracting foreground points. At the same time, it retains the core ideas of previous methods. This cross-category unified model aims to enhance generality and efficiency while maintaining strong tracking performance.

As shown in Fig. 3.15, P2P [146] focuses on directly inferring the target's relative motion between consecutive frames. It also proposes a novel 3D VOT framework that captures accurate motion cues by performing part-to-part fusion of spatial structures in point clouds.

The P2P framework simplifies the traditional motion paradigm, such as that in M2Track, by eliminating auxiliary modules like segmentation and motion refinement. It directly infers the relative motion $(\Delta x_t, \Delta y_t, \Delta z_t, \Delta\theta_t)$ between frames $t-1$ and t using cropped point clouds $\mathcal{P}_{t-1}^{\text{crop}} = \{p_{t-1}^i\}_{i=1}^N$ and $\mathcal{P}_t^{\text{crop}} = \{p_t^i\}_{i=1}^N$, centered on the previous prediction $\mathcal{B}_{t-1}$. This process is expressed as

$$[\Delta x_t, \Delta y_t, \Delta z_t, \Delta\theta_t] = F_m^{\text{pp}}\left(\mathcal{P}_{t-1}^{\text{crop}}, \mathcal{P}_t^{\text{crop}}, \mathcal{B}_{t-1}\right), \tag{3.42}$$

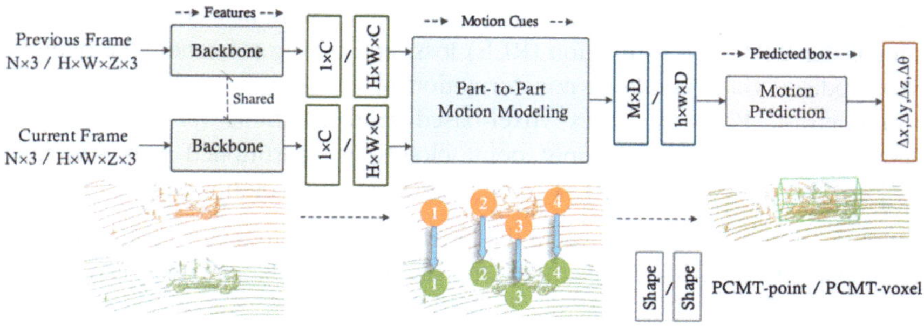

Fig. 3.15 The main architecture diagram of P2P [146]

where F_m^{pp} represents the P2P motion modeling function, consisting of a feature extractor, a part-to-part motion modeling module, and a prediction head. The feature extractor uses a weight-shared network with dual branches to extract spatial structure features from $\mathcal{P}_{t-1}^{\mathrm{crop}}$ and $\mathcal{P}_t^{\mathrm{crop}}$, enabling the distinction between frames. The part-to-part motion modeling module fuses corresponding parts between the spatial structures of the two frames, capturing fine-grained information changes to model robust motion cues. The prediction head then infers the 4-DOF relative motion.

The framework is implemented in two variants: P2P-point and P2P-voxel, which differ in their feature representations. In P2P-point, the feature extractor employs a modified PointNet to process the input point clouds $\mathcal{P}_{t-1/t}^{\mathrm{crop}} \in \mathbb{R}^{N\times 3}$, mapping them to features $f_{t-1/t} \in \mathbb{R}^{N\times C}$, followed by a global max-pooling operation to ensure permutation invariance, yielding semantic features $\mathcal{F}_{t-1/t} \in \mathbb{R}^{1\times C}$. The part-to-part motion modeling concatenates these features along the spatial dimension as

$$\mathcal{F}^{\mathrm{pp}} = \mathrm{Concat}([\mathcal{F}_{t-1}; \mathcal{F}_t], \mathrm{dim} = 0), \tag{3.43}$$

resulting in $\mathcal{F}^{\mathrm{pp}} \in \mathbb{R}^{2\times C}$, where each channel implicitly represents a part. Motion cues are modeled iteratively using cascaded 1D convolution layers across spatial and channel dimensions, expressed as

$$\mathcal{F}_{\mathrm{fusion}}^{\mathrm{pp}} = [\mathrm{Conv1D}([\mathrm{Conv1D}(\mathcal{F}^{\mathrm{pp}})]^\top)]^\top, \tag{3.44}$$

producing $\mathcal{F}_{\mathrm{fusion}}^{\mathrm{pp}} \in \mathbb{R}^{D\times C}$. This allows P2P-point to capture information changes in each part, distinguishing distractors through inconsistency in changes. The prediction head applies max-pooling and a multilayer perceptron (MLP) to infer the relative motion:

$$\mathcal{M}_{t-1,t} = \mathrm{MLP}(\mathrm{Maxpooling}(\mathcal{F}_{\mathrm{fusion}}^{\mathrm{pp}})), \tag{3.45}$$

where $\mathcal{M}_{t-1,t} = (\Delta x_t, \Delta y_t, \Delta z_t, \Delta\theta_t)$. The model is trained end-to-end using a residual log-likelihood estimation (RLE) loss, and during inference, the bounding box is updated via a rigid body transformation $\mathcal{B}_t = F_{\text{trans}}(\mathcal{B}_{t-1}, \mathcal{M}_{t-1,t})$.

In contrast, P2P-voxel uses voxel-based representations for explicit spatial structure preservation. The input point clouds are partitioned into 3D voxels $\mathcal{V}^{\text{crop}}_{t-1/t} \in \mathbb{R}^{W\times L\times H}$, processed by three-stage sparse convolution blocks to produce features $\mathcal{F}_{t-1/t} \in \mathbb{R}^{\frac{W}{s}\times\frac{L}{s}\times D}$, where the height is fused into the channel dimension. Part-to-part fusion concatenates these features along the channel dimension:

$$\mathcal{F}^{\text{pp}} = \text{Concat}([\mathcal{F}_{t-1}; \mathcal{F}_t], \text{dim} = 2), \tag{3.46}$$

yielding $\mathcal{F}^{\text{pp}} \in \mathbb{R}^{\frac{W}{s}\times\frac{L}{s}\times 2D}$, where each spatial position explicitly represents a part. Motion modeling employs 2D convolution blocks to encode spatial and channel changes, simplifying the process compared to P2P-point. The prediction head flattens the features and uses a linear layer followed by an MLP:

$$\mathcal{M}_{t-1,t} = \text{MLP}(\text{Linear}(\text{Flatten}(\mathcal{F}^{\text{pp}}_{\text{fusion}}))). \tag{3.47}$$

P2P-voxel benefits from explicit part-to-part fusion, which allows for precise alignment of corresponding parts. This improves motion modeling accuracy compared to the implicit approach used in P2P-point.

The P2P framework's effectiveness lies in its part-to-part motion modeling, which contrasts with M²Track's approach of fusing all information parts globally. P2P models motion cues part-wise, defined as $F^{\text{pp}}_m(o^1_{t-1,t}, \ldots, o^n_{t-1,t})$, where n is the number of parts (C for P2P-point, $W \cdot L$ for P2P-voxel), compared to M²Track's $F_m(o^1_{t-1}, \ldots, o^n_{t-1}, o^1_t, \ldots, o^n_t)$.

III One-Stream Transformers Methods

Refer to Chap. 2, One-stream Transformers methods adopt a single Transformer architecture to jointly encode features from template and search regions. This approach integrates feature extraction and matching into an end-to-end framework, simplifying the overall system structure while maintaining consistency with the input format of pretraining tasks. This design not only enhances the simplicity and consistency of the system but also improves processing efficiency and tracking performance.

As shown in Fig. 3.16, CorpNet [189] departs from the conventional extractor–matcher–decoder paradigm by proposing a unified encoder and a motion-factorized decoder, aiming to learn target-aware representations more effectively. The method begins with a dynamic 3D sequence of T point clouds $\{\mathcal{P}_i\}_{i=1}^T$ and an initial bounding box (BBox) $\mathcal{B}_1 = \{x_1, y_1, z_1, w_1, l_1, h_1, \theta_1\}$ of a target, where $\{x, y, z\}$ denotes the center coordinates, $\{w, l, h\}$ represents the size, and θ is the heading angle, with the assumption that the BBox size remains constant across frames, simplifying $\mathcal{B}_i$ from $\mathbb{R}^7$ to $\mathbb{R}^4$. The input comprises a template area $\mathcal{P}^t = \{\mathbf{P}^t_i\}_{i=1}^{N_t}$

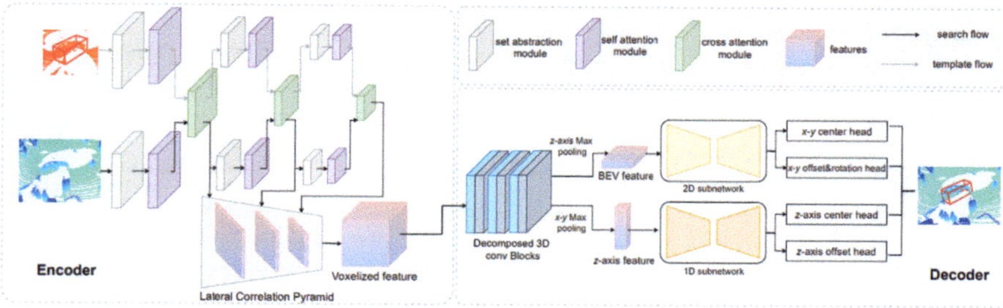

Fig. 3.16 The main architecture diagram of CorpNet [189]

and a search area $\mathcal{P}^s = \{\mathbf{P}_i^s\}_{i=1}^{N_s}$, where $\mathbf{P}_i \in \mathbb{R}^{3+d}$ includes 3D coordinates and d-dimensional features, and N_t and N_s are the respective point numbers.

The encoder of CorpNet integrates multi-level self attentions (SA) and cross attentions (CA) within a three-stage architecture. Each stage featuring a set abstraction module to reduce point numbers, an SA module to enrich feature representations, and a CA module to facilitate fusion and interaction. The SA module employs a local vectorized self-attention mechanism to reduce computational complexity, formalized as follows: for a point with feature $\mathbf{x}_i$ and location $\mathbf{p}_i$, the process involves a linear transformation $\overline{\mathbf{x}}_i = f_1(\mathbf{x}_i)$, attention computation via $\hat{\mathbf{x}}_i = \sum_{j=1}^{k} \sigma\left(\frac{1}{\sqrt{k}} f_a(\mathbf{q}_i - \mathbf{k}_{ij} + \mathbf{e}_{ij})\right) \odot (\mathbf{v}_{ij} + \mathbf{e}_{ij})$, where $\mathbf{q}_i = f_q(\overline{\mathbf{x}}_i)$, $\mathbf{k}_i = f_k(\text{KNN}(\mathbf{p}_i), \overline{\mathbf{x}}_i)$, $\mathbf{v}_i = f_v(\text{KNN}(\mathbf{p}_i), \overline{\mathbf{x}}_i)$, $\mathbf{e}_{ij} = f_e(\mathbf{p}_i - \mathbf{p}_j)$, k is the number of nearest neighbors, f_a is a two-layer MLP, $\odot$ is element-wise multiplication, and σ is a softmax function, with the output $\mathbf{y}_i = f_2(\overline{\mathbf{x}}_i) + \mathbf{x}_i$. The CA module extends this by cross-computing $\mathbf{q}_i$ and $(\mathbf{k}_i, \mathbf{v}_i)$ between template and search regions, yielding $\mathbf{y}_i^s = f_2(\overline{\mathbf{x}}_i^{st}) + \mathbf{x}_i^s$ and $\mathbf{y}_i^t = f_2(\overline{\mathbf{x}}_i^{ts}) + \mathbf{x}_i^t$, enhancing target-relevant features across stages.

To address the sparsity exacerbated by downsampling, CorpNet introduces a lateral correlation pyramid structure within the encoder, fusing hierarchical correlated features from all stages. For each stage's feature $\mathbf{F}^k \in \mathbb{R}^{N_k \times C_k}$, $k \in \{1, 2, 3\}$, a 1D convolution block unifies the feature dimension, and the features are concatenated along the point dimension to form $\mathbf{F} \in \mathbb{R}^{(N_1+N_2+N_3)\times C}$. This is voxelized into a volumetric representation $\mathbf{F}^m \in \mathbb{R}^{C\times H\times L\times W}$ by averaging points within voxel bins, with resolutions defined as:

$$W = \left\lfloor \frac{x_{\max} - x_{\min}}{v_x} \right\rfloor + 1, \tag{3.48}$$

$$L = \left\lfloor \frac{y_{\max} - y_{\min}}{v_y} \right\rfloor + 1, \tag{3.49}$$

$$H = \left\lfloor \frac{z_{\max} - z_{\min}}{v_z} \right\rfloor + 1, \tag{3.50}$$

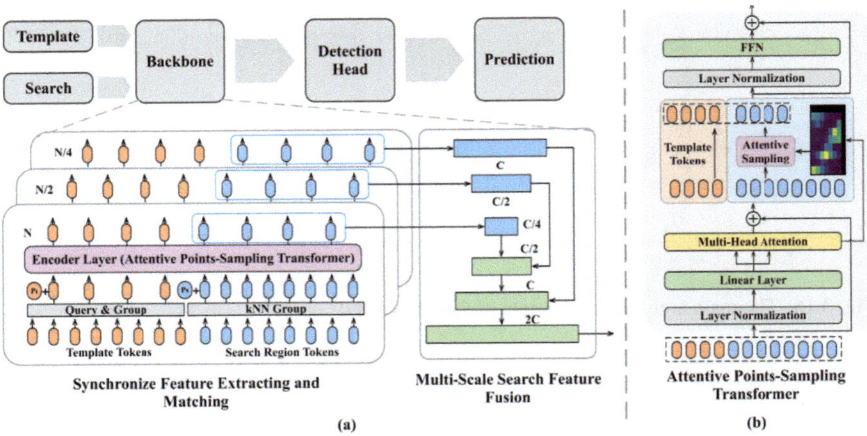

Fig. 3.17 The main architecture diagram of SyncTrack [135]

where $\lfloor\cdot\rfloor$ is the floor operation, and (v_x, v_y, v_z) is the voxel size. The decoder processes $\mathbf{F}^m$ using decomposed 3D convolution blocks, separating motions into 2D BEV and 1D vertical components, with features extracted as $\mathbf{F}^{BEV} = \text{MaxPool}_z\phi(\mathbf{F}^m)$ and $\mathbf{F}^z = \text{MaxPool}_{xy}\phi(\mathbf{F}^m)$, where ϕ represents the convolution blocks. Two subnetworks then predict center classifications and offsets, with the loss function:

$$\mathcal{L} = \lambda_{cls}\left(\mathcal{L}_{cls}^{BEV} + \mathcal{L}_{cls}^{z}\right) + \lambda_{reg}\left(\mathcal{L}_{reg}^{BEV} + \mathcal{L}_{reg}^{z}\right), \tag{3.51}$$

where $\mathcal{L}_{cls}$ uses focal loss and $\mathcal{L}_{reg}$ uses L1 loss, with $\lambda_{cls} = \lambda_{reg} = 1$. The BEV center $\langle c_x, c_y\rangle$ is computed as $c_x = \frac{x - x_{\min}}{v_x}$ and $c_y = \frac{y - y_{\min}}{v_y}$, with discrete values $\lfloor c_x\rfloor$ and $\lfloor c_y\rfloor$.

In contrast, as shown in Fig. 3.17, SyncTrack [135] synchronizes the processes of feature extraction and matching networks by adopting a self-attention mechanism while avoiding the forward propagation of Siamese networks, thereby simplifying the model design into a single-branch and single-stage approach. Given a sequence of point clouds $\{P_i\}_{i=1}^{T}$ (where T is the number of points in each frame) and an initial bounding box B_1 of the target, the goal of SOT is to localize the target bounding boxes $\{B_i\}_{i=2}^{T}$ in all frames online. Following this approach, a template point cloud $P_t = \{pt_i\}_{i=1}^{N_t}$ and a search region $P_s = \{ps_i\}_{i=1}^{Ns}$ are generated, where N_t and Ns represent the number of template and search region points, respectively. The template P_t is generated by cropping and centering the target in the initial frame based on the initial bounding box.

To achieve effective feature extraction and matching, SyncTrack leverages Transformers' long-range relation-modeling capability. Specifically, it utilizes self-attention modules to build a single-branch backbone. In this framework, a Query & Group module is employed to sample template seeds denoted as $P_t \in R^{N_t \times C}$ using the farthest point sampling (FPS) method before jointly forwarding. This module also groups the k-nearest points to aggregate features. For the search region point cloud, represented as $P_s \in R^{Ns \times C}$, only the Group module is used to aggregate neighbor information without reducing the number of search points. Subsequently, the template and search seeds are concatenated with a joint parametric positional embedding for token localization. This process can be expressed as:

$$T_t = kNN(FPS(P_t)), P_t \in R^{N_t' \times C'}, N_t' < N_t, \tag{3.52}$$

$$T_s = kNN(P_s), P_s \in R^{Ns \times C'}, \tag{3.53}$$

$$T_{ts} = [T_t; T_s] + pe, T_{ts} \in R^{(N_t + Ns) \times C'}. \tag{3.54}$$

Afterward, linear layers project the input tokens into query, key, and value latent representations, and the head-wise joint attention map is calculated to model intra- and inter-relations of template and search tokens:

$$Q_{ts}, K_{ts}, V_{ts} = W_q(T_{ts}), W_k(T_{ts}), W_v(T_{ts}), \tag{3.55}$$

$$A_{tms} = Softmax\left(\frac{Q_{tms}(K_{ts})^{\top}}{\sqrt{C'}}\right). \tag{3.56}$$

Based on the joint head-wise attention map A_{tms} of template and search tokens, the features extracted by multi-head attention are:

$$T_{ts}' = [A_{t1s}V_{ts1}, A_{t2s}V_{ts2}, \ldots, A_{tMs}V_{tsM}]W + T_{ts}, \tag{3.57}$$

where M is the number of attention heads and W is the weight of a multilayer perceptron (MLP).

The single-branch structure enables synchronization of the processes of feature extraction and matching. The formula of joint attention A_{tms} in Eq. 3.56 can be expanded as follows (σ represents the softmax operation):

$$[\sigma Q_{tm}(K_{mt})^{\top}/\sqrt{C'}, \sigma Q_{tm}(K_{ms})^{\top}/\sqrt{C'}; \sigma Q_{sm}(K_{mt})^{\top}/\sqrt{C'}, \sigma Q_{sm}(K_{ms})^{\top}/\sqrt{C'}]. \tag{3.58}$$

From this, the newly extracted search region features can be obtained:

$$T_{sm}' = \sigma Q_{sm}(K_{mt})^{\top}/\sqrt{C'}Vm_t + \sigma Q_{sm}(K_{ms})^{\top}/\sqrt{C'}V_{ms}, \tag{3.59}$$

where $[V_{mt}; V_{ms}] = V_{tsm}$.

Through these steps, SyncTrack achieves efficient and accurate 3D VOT. It does so by synchronizing feature extraction and matching within a streamlined single-branch framework. This design significantly outperforms previous state-of-the-art methods on various datasets.

3.3.3 LiDAR-Based VOT Methods Performance Comparison

The study of 3D object tracking in point clouds is primarily conducted on well-known datasets such as KITTI and nuScenes, which provide essential benchmarks for evaluating and advancing three-dimensional target tracking algorithms.

The performance of LiDAR-based tracking methods is presented in Table 3.3, along with additional results provided here in Table 3.4. It should be noted that the evaluation on the nuScenes dataset involves two distinct configurations. The first configuration involves training the model on the nuScenes training split, considering object categories such as Car, Pedestrian, Truck, Trailer, and Bus—corresponding to the row labeled "nuScenes1" in Table 3.4. The second configuration evaluates performance using models pretrained on the KITTI dataset, applied directly to the nuScenes test set without any further training on nuScenes data; this is represented in the row labeled "nuScenes2" in Table 3.4. With the passage of time, the overall performance of LiDAR-based trackers on KITTI and nuScenes has been continuously improving. For example, from SC3D in 2019 to MTMTrack in 2023, the performance has more than doubled, demonstrating significant progress and continuous development in this field.

3.3.4 Summary and Challenges

For 3D object tracking based on LiDAR, the three main methods include Siamese networks, motion-model-based approaches, and one-stream methods. These methods fully leverage the unique advantages of point cloud data, such as high-precision 3D spatial representation, rich geometric information, and robustness in complex environments. Siamese networks achieve accurate target matching through a dual-branch structure, capable of handling complex geometric transformations and adapting to varying environmental conditions. Motion-model-based approaches utilize historical data to predict target positions, enhancing the ability to forecast motion trajectories, making them particularly suitable for addressing rapid target movements and occlusion issues. Meanwhile, one-stream methods integrate feature extraction and motion estimation into a unified framework, providing a simplified and efficient solution while maintaining invariance to rotation and translation. These three methods each address different challenges and collectively promote the development of LiDAR-based object tracking technology, demonstrating strong application potential in fields such as autonomous driving and robotic navigation.

However, the use of point cloud modality still encounters several challenges:

1. **Sparsity and Irregularity**: Point cloud data, acquired from sensors like LiDAR, consist of discrete points that are often sparse and irregularly distributed. Unlike traditional image data, where pixels are uniformly arranged, the points in a point cloud are not consistently spaced, which makes their processing and analysis more challenging. Specifically, point clouds are composed of scattered data

Table 3.3 Comparison of the performance of different LiDAR-based trackers on the KITTI dataset

Paradigms	Trackers	Publication	KITTI				
			Car(6424)	Pedestrian(6088)	Van(1248)	Cyclist(308)	Mean(14068)
Siamese network	SC3D [56]	CVPR2019	41.3 / 57.9	18.2 / 37.8	40.4 / 47.0	41.5 / 70.4	31.2 / 48.5
	P2B [153]	CVPR2020	56.2 / 72.8	28.6 / 49.6	40.8 / 48.4	32.2 / 44.7	42.4 / 60.0
	MLVSNet [201]	ICCV2021	56.1 / 74.0	34.1 / 61.1	52.0 / 61.4	34.3 / 44.5	45.7 / 66.6
	BAT [259]	ICCV2021	60.5 / 77.7	42.1 / 70.1	52.3 / 67.0	33.7 / 45.4	51.2 / 72.8
	V2B [86]	NeurIPS2021	70.5 / 81.4	48.3 / 73.5	50.1 / 58.1	40.8 / 49.8	58.4 / 75.2
	STNet [85]	ECCV2022	72.1 / 84.0	49.9 / 77.2	58.0 / 70.5	73.5 / 93.7	61.3 / 80.0
	GLT-T [144]	AAAI2023	68.2 / 82.1	52.4 / 78.8	52.6 / 62.9	68.9 / 92.1	60.1 / 79.3
	OSP2B [145]	IJCAI2023	67.5 / 82.3	53.6 / 85.1	56.3 / 66.2	65.6 / 90.5	60.5 / 82.3
	CXTrack [212]	CVPR2023	69.1 / 81.6	67.0 / 91.5	60.0 / 71.8	74.2 / 94.3	67.5 / 85.3
	M3SOT [120]	None2023	75.9 / 87.4	66.6 / 92.5	59.4 / 74.7	70.3 / 93.4	70.3 / 88.6
Motion modeling	M2Track [258]	CVPR2022	65.5 / 80.8	61.5 / 88.2	53.8 / 70.7	73.2 / 93.5	62.9 / 83.4
	StreamTrack [131]	None2023	72.6 / 83.7	70.5 / 94.7	61.0 / 76.9	78.1 / 94.6	70.8 / 88.1
	MTMTrack [112]	RAL2023	73.1 / 84.5	70.4 / 95.1	60.8 / 74.2	76.7 / 94.5	70.9 / 88.4
	MoCUT [147]	ICLR2024	67.6 / 80.4	63.3 / 90.0	64.5 / 78.8	76.7 / 94.2	65.8 / 85.0
One stream	CorpNet [189]	CVPRW2023	73.6 / 84.0	55.6 / 82.4	58.7 / 66.4	74.3 / 94.2	64.5 / 82.0
	SyncTrack [135]	ICCV2023	73.3 / 85.0	54.7 / 80.5	60.3 / 70.0	73.1 / 93.8	64.1 / 81.9

Table 3.4 Comparison of the performance of different LiDAR-based trackers on nuScenes dataset

	nuScenes1						nuScenes2				
Trackers	Car(64159)	Pedestrian(3322)	Truck(13587)	Trailer(3352)	Bus(2953)	Mean(117278)	Car(15578)	Pedestrian(8019)	Truck(3710)	Bicycle(501)	Mean(27808)
SC3D [56]	22.31 / 21.93	11.29 / 12.65	30.67 / 27.73	35.28 / 28.12	29.35 / 24.08	20.70 / 20.20	25.0 / 27.1	14.2 / 16.2	25.7 / 21.9	17.0 / 18.2	21.8 / 23.1
P2B [153]	38.81 / 43.18	28.39 / 52.24	42.95 / 41.59	48.96 / 40.05	32.95 / 27.41	36.48 / 45.08	27.0 / 29.2	15.9 / 22.0	21.5 / 16.2	20.0 / 26.4	22.9 / 25.3
BAT [259]	40.73 / 43.29	28.83 / 53.32	45.34 / 42.58	52.59 / 44.89	35.44 / 28.01	38.10 / 45.71	22.5 / 24.1	17.3 / 24.5	19.3 / 15.8	17.0 / 18.8	20.5 / 23.0
V2B [86]	–	–	–	–	–	–	31.3 / 35.1	17.3 / 23.4	21.7 / 16.7	22.2 / 19.1	25.8 / 29.0
STNet [85]	–	–	–	–	–	–	32.2 / 36.1	19.1 / 27.2	22.3 / 16.8	21.2 / 29.2	26.9 / 30.8
CXTrack [212]	–	–	–	–	–	–	29.6 / 33.4	20.4 / 32.9	27.6 / 20.8	18.5 / 26.8	26.5 / 31.5
M3SOT [120]	–	–	–	–	–	–	34.9 / 39.9	23.3 / 25.6	30.4 / 27.0	16.5 / 22.6	30.6 / 33.7
M2Track [258]	55.85 / 65.09	32.10 / 60.92	57.36 / 59.54	57.61 / 58.26	51.39 / 51.44	49.23 / 62.73	–	–	–	–	–
StreamTrack [131]	62.05 / 70.81	38.43 / 68.58	64.67 / 66.60	66.67 / 64.27	60.66 / 59.74	55.75 / 69.22	–	–	–	–	–
MTMTrack [112]	57.05 / 65.94	37.08 / 72.80	59.37 / 63.69	59.73 / 60.44	55.46 / 54.31	51.70 / 67.17	–	–	–	–	–
MoCUT [147]	57.32 / 66.01	33.47 / 63.12	61.75 / 64.38	60.90 / 61.84	57.39 / 56.07	51.19 / 64.63	–	–	–	–	–
CorpNet [189]	–	–	–	–	–	–	35.0 / 38.4	21.3 / 33.6	39.7 / 36.3	26.9 / 43.5	31.8 / 36.8
SyncTrack [135]	–	–	–	–	–	–	36.7 / 38.1	19.1 / 27.8	39.4 / 38.6	23.8 / 30.4	31.8 / 35.1

points that lack continuity in space, resulting in sparse information. This sparsity complicates the application of traditional grid-based methods, which rely on regular, dense data structures. Furthermore, the irregular distribution of points means that there is no consistent spatial pattern, making tasks like interpolation, resampling, and densification particularly difficult. Ensuring geometric accuracy while accounting for the relationships between points adds an extra layer of complexity.

2. **Multi-modal Data Fusion**: In many applications, point cloud data are often combined with data from other sensors, such as RGB images or radar data. Due to the differences in structure and characteristics between point clouds and other modalities (e.g., images), effectively fusing data from different modalities to extract complementary information and improve system performance remains a challenge.
3. **High Computational Complexity and Storage Requirements**: Point cloud data consist of a large number of 3D data points. In large-scale scenes or high-precision tasks, this can lead to a significant increase in computational complexity and memory requirements. For real-time applications such as autonomous driving, it is crucial to efficiently process vast amounts of point cloud data within limited computational resources. Optimizing the storage, transmission, and processing of point cloud data to reduce computational costs remains an important research topic in the field.
4. **Processing in Dynamic Scenes**: In dynamic environments, point cloud data are often affected by target motion, occlusions, or environmental changes. Achieving stable object tracking and detection in such scenarios, especially in complex urban environments or on highways, remains a challenging problem.

3.4 Event-Based VOT

3.4.1 Modal Characteristics and Application Scenarios

A new type of neuromorphic vision sensor has emerged. It is inspired by biological visual systems, especially the retinal neural structure. This sensor creates a new bio-inspired paradigm for visual information acquisition [18]. The core principle [82] is that each pixel independently detects brightness changes [42]. An "event" is generated when a pixel's brightness change exceeds a set threshold. This event is output asynchronously. It includes a timestamp, pixel coordinates, and the polarity of the brightness change (brightening or darkening).

Advantages [49] of cameras based on this new paradigm include:

1. **High Temporal Resolution**: Event cameras offer high temporal resolution at the microsecond level. This far surpasses traditional frame-based cameras. Traditional cameras are usually limited by frame rates like 30fps or 60fps. This high temporal resolution allows event cameras to capture fast dynamic changes.

They can still acquire high-quality data during high-speed motion [40]. This feature is crucial for real-time dynamic scene processing. It benefits applications like drone visual feedback during high-speed flight. It also helps robots avoid obstacles in complex environments.
2. **Low Latency**: Each pixel in an event camera operates independently without waiting for a global exposure time, enabling asynchronous output of event data. This mechanism greatly reduces output latency, often to sub-millisecond levels. It also sharply lowers bandwidth needs for data transmission and processing [41, 54]. Unlike traditional cameras, event cameras avoid transmitting redundant background data and output only event-based information related to brightness changes. This efficient data processing enables event cameras to respond quickly in dynamic scenes. It makes them ideal for applications with high real-time demands. Typical examples include autonomous driving, robot navigation, and augmented reality (AR) systems.
3. **Low Power Consumption**: Event cameras output data only when brightness changes exceed a preset threshold. This avoids transmitting and processing redundant data. Traditional cameras often generate such unnecessary information. As a result, event cameras consume very low power [4], typically around 10 milliwatts, with some prototypes even operating below 10 milliwatts. This low-power characteristic makes them particularly suitable for battery-powered devices and embedded systems, such as drones, IoT devices, and robots requiring long operational durations. In resource-constrained environments, the low-power nature of event cameras significantly extends device runtime while reducing energy consumption.
4. **High Dynamic Range**: Event cameras feature a dynamic range of up to 140 dB, surpassing the 60–80 dB range of traditional cameras. This allows them to capture clear images in extreme lighting conditions, such as direct sunlight or very dark environments, while preserving details in both bright and dark areas. This high dynamic range ensures strong performance in high-contrast scenes. It works well in outdoor sunlight and dim indoor lighting. As a result, it provides reliable input for visual tasks.

In summary, with the reduction in manufacturing costs of event cameras, improved availability of simulators/emulators [60, 61, 89, 117, 158, 256] and datasets [12, 16, 67, 79, 80, 121, 130, 139, 140, 161, 195], and the rapid advancement of deep learning, both academia and industry have shown strong interest [18] in this new paradigm of cameras. Event-based technologies have spread rapidly, and the number of works in this field has further expanded. Typically, the event modality can be applied in:

- Feature Detection and Tracking [3, 37, 78, 94]: Feature detection and tracking are two core tasks in computer vision, providing support for higher-level visual tasks. They are widely used in object recognition, localization, tracking, visual odometry (VO), simultaneous localization and mapping (SLAM), and more. Feature detection identifies key points or salient regions in images or videos. These points are often called "feature points" or "interest points". These are pixel

positions with unique appearances, such as corners, edges, and blobs. Feature tracking means locating and matching feature points across frames. It works on consecutive images or video sequences. The goal is to find point positions in later frames. This enables real-time object tracking. These technologies make full use of event properties. Key features include high temporal resolution and sparsity. As a result, they perform well in dynamic scenes and complex environments.
- Corner Detection and Tracking [2, 58, 105, 136, 141, 173]: Corner detection and tracking are two core technologies in computer vision, widely used in fields such as robot navigation, object tracking, augmented reality, and video analysis. Corner detection is a technique for extracting points with distinct geometric features from images or event streams. Corners are points where two edges meet in an image and typically exhibit significant brightness changes, making them crucial feature points in visual tasks. Corner tracking is the process of dynamically locating and identifying changes in the position of corners in consecutive images or event streams. It achieves this by associating corners across frames or event streams to form corner trajectories. The event modality has several advantages. It offers high temporal resolution, low power use, and sparse, asynchronous data. These traits support accurate and robust corner detection and tracking. They also suit efficient use on embedded and edge devices.
- Simultaneous Localization And Mapping (SLAM) [68, 274]: SLAM is a key technology for robots and intelligent systems. It allows them to locate themselves in unknown environments. At the same time, it helps build a map of the surroundings. SLAM answers two key real-time questions: "Where am I?" and "What does the environment look like?" It uses sensor data to solve these problems. This technology plays a vital role in robot navigation, AR, VR, and autonomous driving. Event cameras offer high dynamic range and fine time resolution. They also consume little power. These features make them ideal for challenging scenes. They perform well in fast motion, high dynamics, and low-light conditions. These attributes enhance the robustness and computational efficiency of SLAM systems.

In addition, as event-based vision technology continues to evolve, a growing number of downstream tasks—such as depth estimation [268], 3D shape reconstruction [8], object segmentation [27], image deblurring [149, 156], action recognition [52, 53, 119], and VOT (which will be detailed in this section)—are being incorporated into the scope of event modality applications.

3.4.2 *Inductive Classification and Typical Methods*

In the field of VOT, event-based methods are relatively new compared to traditional RGB-based approaches and have received less attention so far. However, as the unique advantages of event cameras in dynamic scenes are gradually uncovered, event-based VOT is emerging as an area of growing interest among researchers.

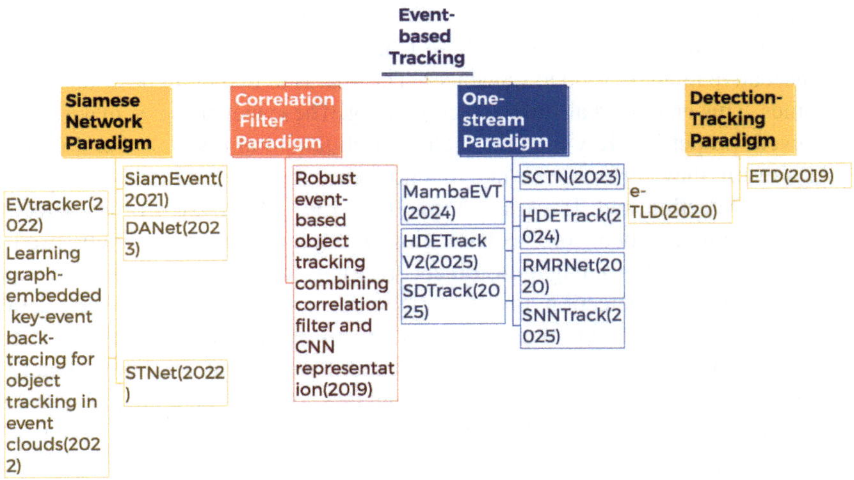

Fig. 3.18 Classification of VOT Method Based on Event

Following the classification framework introduced in Chap. 2, as shown in Fig. 3.18, event-based VOT methods can be categorized into the following four major paradigms: Siamese network, correlation filter, one-stream network, and detection-tracking paradigms. It is particularly noteworthy that while Siamese networks and correlation filters dominate RGB-based VOT, they remain underexplored in the event-based context. In contrast, one-stream paradigm has gained significant attention in recent years and is now becoming research hotspot in event-based VOT.

Next, this section will provide an in-depth analysis of each paradigm, covering its overview, research priorities, and key methods.

3.4.2.1 Siamese Network Paradigm

Chapter 2 and Sect. 3.1.2.1 have introduced the Siamese network paradigm, so no further elaboration will be provided here. Next, the focus will be on specific tracking methods directly.

SiamEvent [19] first embeds event data into a shared feature extractor. It then uses a Siamese network to compute similarity. The comparison is made between the target and the search region. The core of this method is edge-aware similarity learning. It focuses on matching edge structures. The goal is to find the part most similar to the target. This is done by learning the similarity of target edges. To prevent tracking drift, SiamEvent also introduces target edge initialization and an edge detector. Edge initialization generates a clear target edge structure, while the edge detector pauses tracking in static scenes to save computational resources. Experiments show that SiamEvent demonstrates higher robustness under HDR and motion blur conditions.

EVtracker [249] dynamically adjusts the spatiotemporal domain of event data through online learning to ensure stable long-term tracking. It first evaluates the feature quality of event data to dynamically adjust the spatiotemporal domain and generate high-quality event images. Then, it uses a Siamese network for feature extraction and target matching, with classification and regression modules for tracking. In case of tracking failure, a global detection mechanism reinitializes the tracker to address target loss in long-term tracking. Experiments show that EVtracker achieves stable and accurate tracking under fast motion, rotation, and complex lighting conditions.

DANet [46] adopts a Siamese network architecture enhanced with a Transformer. It discovers and identifies target objects by leveraging motion cues and object contours simultaneously. The framework consists of two subnetworks: a motion-aware network that extracts dynamic scene features and an object-aware network that identifies the target through correlations between template and search frames. The Transformer uses a multi-head self-attention mechanism to capture global contrasts between dynamic objects and the background, thereby enhancing object embedding. Experiments demonstrate that DANet achieves superior tracking performance under conditions such as camera motion, multiple-object motion, and varying illumination.

The method [272] introduces a density-insensitive downsampling strategy to sample a subset called key-events. Then, it uses a graph-based network to process key events. This network embeds irregular spatiotemporal data into a high-dimensional space. It then predicts target likelihood using semantic-driven Siamese matching. Additionally, it proposes motion-aware target likelihood prediction. This method traces back the initial positions of key events. It then uses motion flow to estimate their likelihood of being targets. Finally, it obtains the final bounding box by adaptively fusing two intermediate ones.

STNet [245] dynamically extracts and fuses information from both temporal and spatial domains. It uses a Transformer-based module to capture global spatial information and an SNN-based module to extract temporal cues. The spiking threshold of the SNN module is dynamically adjusted based on statistical cues from spatial information, which is crucial for robust feature extraction. Finally, a novel cross-domain attention algorithm dynamically fuses the two feature branches. Experiments show that STNet outperforms existing methods in both accuracy and speed across three event-based datasets.

3.4.2.2 Correlation Filter Paradigm

Chapter 2 and Sect. 3.1.2.2 have introduced the correlation filter paradigm, so no further elaboration will be provided here. Specific methods will be elaborated in Sect. 3.4.2.5.

3.4.2.3 One-Stream Paradigm

Chapter 2 and Sect. 3.1.2.3 have introduced the one-stream paradigm, so no further elaboration will be provided here. Next, the focus will be on specific tracking methods directly.

SCTN [87] employs a deep convolutional spiking neural network (SNN) to directly process event streams. By introducing an exponential Intersection over Union (IoU) loss function, it fully leverages the temporal information of events while maintaining sparse representations. The SCTN architecture consists of three convolutional layers and three fully connected layers, enabling direct processing of event streams without additional preprocessing. Additionally, SCTN introduces a new tracking dataset for event cameras. It is called DVSOT21 and includes nine sequences. These sequences cover challenges like fast motion, rich textures, and partial occlusion. Experimental results show that SCTN performs well in handling challenges like partial occlusion, fast motion, and background clutter. Notably, it accurately captures events generated by target contours under partial occlusion, achieving more robust tracking. Moreover, SCTN demonstrates significant advantages in energy efficiency.

MambaEVT [196] uses a state space model as the backbone network to reduce computational complexity and enhance tracking performance. It takes the search region and target template as inputs. These inputs are processed through the Mamba network. The network performs feature extraction and interaction learning. The output features are then used for target localization. Additionally, MambaEVT introduces a dynamic template update strategy. This strategy employs a Memory Mamba network to generate dynamic templates that adapt to changes in the target's appearance. By maintaining a template library that combines long-term and short-term memory, it selectively updates templates to better accommodate changes in the target's appearance. Experimental results show that this method is particularly effective in dealing with challenges such as cluttered backgrounds and changes in target appearance. Especially in low-light and fast-motion scenarios, the dynamic template update strategy significantly improves the robustness and accuracy of tracking.

HDETrack [195] is based on hierarchical knowledge distillation, enhancing tracking performance through knowledge transfer from multi-modal or multi-view data. It first trains a teacher Transformer network using RGB frames and event streams. Then, it guides the learning of a student Transformer network via knowledge distillation strategies, including similarity matrices, feature representations, and response maps. To solve the problem of low-resolution datasets, HDETrack introduces a new dataset. It is called EventVOT and contains 1141 high-resolution videos. These videos cover diverse target categories and challenging factors. Experiments show that it excels in handling challenges like target deformation, camera motion, and similar-object interference. Especially in low-light and fast-motion scenes, the hierarchical knowledge distillation strategy significantly boosts tracking robustness and accuracy.

HDETrack V2 [194], building on HDETrack, introduces Test-Time Tuning (TTT) and Adaptive Search Region (ASR) strategies to enhance the model's generalization and robustness. Experimental results show that it performs well in handling challenges like target deformation, camera motion, and background clutter. Particularly in fast-motion and partial-occlusion scenarios, the TTT and ASR strategies further boost tracking performance.

RMRNet [22] introduces a new representation called TSLTD. It stands for Time-Surface with Linear Time Decay. This method encodes spatial and temporal data from asynchronous retinal events. The resulting TSLTD frames clearly show motion patterns. These frames are fed into RMRNet for end-to-end 5-DoF object motion regression. Experiments show that this method outperforms existing traditional and event-based camera object tracking methods in challenging environments like fast motion and low illumination.

SDTrack [163] proposes a Global Trajectory Prompt (GTP) method. It captures global trajectory information from event flows. This information is then aggregated into event images. The goal is to enhance spatio-temporal representation. SDTrack includes a spiking MetaFormer backbone [228, 229]. It also has a simple tracking head. The model directly predicts normalized coordinates using spike signals. Experiments show that SDTrack achieves state-of-the-art performance in multiple event-based tracking benchmarks while maintaining the lowest parameter count and energy consumption.

SNNTrack [244] proposes a brain-inspired adaptive Leaky Integrate-and-Fire neuron that can adjust the membrane time constant based on inputs. This accelerates the decay of noise features and reduces the loss of valuable information. It also introduces a temporal feature aggregator to enhance network adaptability by assigning attention weights across the time dimension. Experiments confirm its effectiveness on multiple event-based tracking datasets and show its generalization ability for event classification tasks.

3.4.2.4 Detection-Tracking Paradigm

Section 3.1.2.5 has introduced the detection-tracking paradigm, so no further elaboration will be provided here. Next, the focus will be on specific tracking methods directly.

ETD [21] converts asynchronous events into frame representations. These events are captured by an event camera. The method produces Adaptive Time-Surface with Linear Time Decay (ATSLTD) frames. These frames effectively record the contours of target objects. The event camera captures asynchronous events. These events are transformed into ATSLTD frames. The frames clearly show the contours of target objects. They also provide high temporal resolution. Based on these ATSLTD frames, the method employs a two-stage detection-tracking framework: first, the detection stage uses the EdgeBoxes [273] detector to generate candidate bounding boxes for the target object; then, the tracking stage uses an IoU-based tracker to select the best bounding box from the candidates as the tracking result for the current frame. The method introduces a new metric called Non-Zero Grid Entropy. It measures the amount of information in ATSLTD frames. The method also uses a confidence interval. This triggers the generation of ATSLTD frames asynchronously. Experimental results show that this method performs exceptionally well in complex environments involving fast motion, low illumination, and high dynamic range scenes.

The e-TLD [157] framework combines a local tracker and a global detector. The local tracker uses an event-based sliding window. It searches within a local region around the target. The tracker also updates the object's representation online. This helps it adapt to appearance changes. When tracking fails, the global detector is activated to search for the object across the entire image using a global sliding window approach. The core idea of e-TLD is to improve object representation. It uses Bayesian bootstrapping to boost discriminative power. The method can also detect and re-track the object. This happens when the object re-enters the field of view. Experimental results show that the framework is effective. It can track and detect objects of different shapes and sizes. It performs well in complex backgrounds. It also remains robust under changing light and occlusion.

3.4.2.5 Typical Methods

I Siamese Network Paradigm Typical Method: DANet [46]

Although some studies [7, 57] have distinguished between background and moving targets using clustering algorithms, these handcrafted strategies are not generalizable across different scenarios. Deep learning–based methods [22, 243, 245] improve feature extraction. They also enhance the discriminability of targets. However, they do not model the link between targets and motion explicitly. As a result, these methods perform poorly in locating the target when there are distractors with similar motion or background noise in the scene. DANet solves these issues effectively. It introduces a motion-aware network and a target-aware network. These networks work together. They use motion cues and object contour information at the same time.

Specifically, DANet mainly consists of a feature extraction network, a Target-Aware Network (TAN), a Motion-Aware Network (MAN), a Target Fusion Module (TFM), and a regression head, as shown in Fig. 3.19. Below is a detailed introduction to each module.

Backbone Network Implementation Details

ResNet-18 [73] serves as the backbone network for the feature extractor to process the template frame and the search frame. The template frame is fed into the ResNet-18 network. The search frame is also processed by the same network. Each input generates feature representations. These features include edge, motion, and background contrast information. These feature maps are then passed to the TAN and MAN.

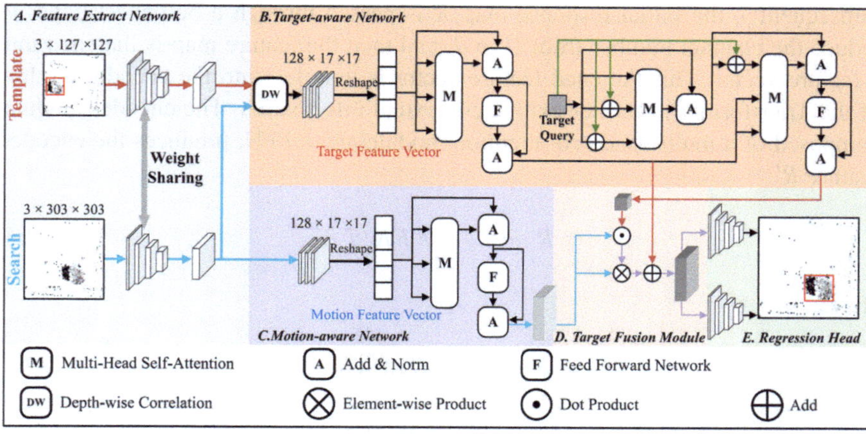

Fig. 3.19 The main architecture diagram of DANet [46]

TAN Implementation Details

TAN identifies and localizes the target object by matching the target features in the template frame and the search frame. Its inputs include the template frame features $f(z)$ and the search frame features $f(x)$ generated by the feature extraction network. Next, the template frame features act as convolutional kernels. They are applied to the search frame features using depthwise separable convolution. This process generates the initial target response map R.

$$R = \psi(z, x) = \phi(f(z)) \star \varphi(f(x)) \tag{3.60}$$

Here, ϕ and φ are two distinct convolutional layers, and $\star$ denotes the depthwise separable convolution operation. The initial response map R represents the similarity between the template frame and the search frame, reflecting the location of the target within the search frame. To improve target identification accuracy, TAN uses a Transformer module. This module refines the initial response map. It applies a multi-head self-attention mechanism. It also uses a feed-forward network (FFN) for further enhancement:

$$MultiHead(Q, K, V) = Concat(H_1, \ldots, H_n)W^O \tag{3.61}$$

Here,

$$H_i = Attention\left(QW_i^Q, KW_i^K, VW_i^V\right) \tag{3.62}$$

$$Attention(Q, K, V) = \mathrm{softmax}(QK^\top/\sqrt{d_k})V \tag{3.63}$$

W^O, W_i^Q, W_i^K, and W_i^V denote model parameters.

Subsequently, the initial response map R is passed through a bottleneck layer to reduce the channel number from C to d, and then the feature map is flattened into a feature vector. This flattened feature vector is then fed into the encoder module of the Transformer to further extract contextual information. The encoder module, composed of a multi-head self-attention module and a FFN, produces the encoded feature R':

$$R' = \mathcal{F}_e + FFN(\mathcal{F}_e) \tag{3.64}$$

Here,

$$\mathcal{F}_e = R + MultiHead(R, R, R) \tag{3.65}$$

$$FFN(x) = \max(0, xW_1 + b_1)W_2 + b_2 \tag{3.66}$$

W and b are weight matrix and basis vectors.

Finally, the encoded feature R' is input into the decoder module to generate the target embedding. The decoder module, also composed of a multi-head self-attention module and an FFN, produces the target embedding T:

$$T = \mathcal{F}_d + FFN(\mathcal{F}_d) \tag{3.67}$$

Here,

$$\mathcal{F}_d = A + MultiHead(A + TQ, R', R') \tag{3.68}$$

$$A = TQ + MultiHead(TQ + TQ, TQ + TQ, TQ) \tag{3.69}$$

TQ is a target query.

MAN Implementation Details

The primary task of the MAN is to extract dynamic information from the scene to identify potential moving objects. It takes the feature map $f(x)$ of the search frame, generated by the feature extraction network, as input. Subsequently, it employs a 3×3 convolutional layer with a stride of 2 to reshape the feature map into M. Following this, it captures the dynamic information within the feature map through the multi-head self-attention mechanism and the FFN:

$$M' = \mathcal{M}_e + FFN(\mathcal{M}_e) \tag{3.70}$$

Here,

$$\mathcal{M}_e = M + MultiHead(M, M, M) \tag{3.71}$$

TFM Implementation Details

The TFM generates a confidence map for target localization by fusing the target embedding produced by the TAN and the motion features produced by the MAN. The TFM calculates the similarity between the two inputs. It then uses the similarity scores to weight the motion features. This enhances the features of the target region. At the same time, it suppresses features of other potential distractors.

Regression Head Implementation Details

The regression head's task is to predict the precise location of the target based on the feature map fused by the TFM. Specifically, it processes the fused feature map through a fully convolutional network to generate the target's center coordinates and bounding box scale information.

Loss Function Implementation Details

Finally, the composite loss function of DANet combines the L1 loss and the IoU loss, controlling their contributions through the balancing weights λ_{iou} and λ_{L_1}:

$$L = \lambda_{\text{iou}} L_{\text{iou}} \left(b_i, \hat{b_i}\right) + \lambda_{L_1} L_1 \left(b_i, \hat{b_i}\right) \tag{3.72}$$

b_i and $\hat{b_i}$ represent the ground-truth and the predicted box.

II Correlation Filter Paradigm Typical Method: [100]

Previous research on event-based object tracking explored several methods. Techniques like continuous event clustering [118] can handle events in real time. Iterative closest point algorithms [143] also support real-time event processing. However, they are prone to errors in the presence of many noise events, as these noise events may be mistakenly classified as part of the target object. Feature-based tracking methods [35, 132] improve robustness in complex scenes. However, they often need to accumulate events within a time window. This step is necessary to extract useful features. As a result, it reduces the real-time performance of tracking. The proposed method [100] combines a correlation filter with CNN feature representation. This approach effectively handles noise events and complex background textures. It also copes well with occlusions and rapid changes in target shape. The algorithm workflow is as follows:

Rate Coding Implementation Details

First, perform rate coding, which represents the pixel value by calculating the number of events generated by each pixel within a specific time window. This method effectively reduces the impact of noise events by integrating the events of each pixel within the time window. The physical significance lies in the fact that the number of events per pixel represents the frequency of intensity changes exceeding the threshold at that pixel. The event rate of each pixel can be expressed as:

$$Rate(t) \approx \frac{TCON(t)}{\theta} = \frac{1}{\theta} \frac{d \ln(I)}{dt} \tag{3.73}$$

Here, $TCON$ represents temporal contrast, $I(t)$ is the photocurrent, and θ is a constant.

Feature Extraction Implementation Details

Next, the rate-coded image is input into a pre-trained VGGNet-16 [165], which uses feature maps from the conv1_1, conv2_2, and conv3_3 layers. The conv1_1 layer extracts low-level features like edges and textures, the conv2_2 layer extracts mid-level features, and the conv3_3 layer extracts high-level semantic features.

Correlation Filter Implementation Details

Subsequently, the method combines multilayer feature maps as input for the correlation filter. The training of this filter can be formulated as a linear least squares problem:

$$w = \arg_{f_t} \min \sum_{h,w} \left\| f_t \cdot x^l_{h,w} - y_{h,w} \right\|^2 + \lambda \, \|ft\|_2^2 \tag{3.74}$$

$x^l_{h,w}$ represents the shifted samples of the feature map, $y_{h,w}$ is the Gaussian label indicating the desired target position, and λ is the regularization parameter used to prevent overfitting. Here,

$$y_{h,w} = \exp\left(-\frac{(h - H/2)^2 + (w - W/2)^2}{2\sigma^2}\right) \tag{3.75}$$

h and w are the row and column indices of the feature map; H and W are the height and width of the feature map; σ is the bandwidth parameter of the Gaussian kernel. The above optimization problem can be efficiently solved in the frequency domain using the Fast Fourier Transform (FFT). The expression for the filter in the frequency domain is:

$$F^c = \frac{Y \odot \overline{X}^c}{\sum_{i=1}^{C} X^i \odot \overline{X}^i + \lambda} \tag{3.76}$$

Here, Y is the Fourier transform of the Gaussian label y, $\odot$ denotes element-wise multiplication, and C is the number of channels in the feature map.

Finally, the correlation response map r^l is computed using an inverse FFT. The position of its maximum value is then located. This position indicates the estimated location of the target object in the current frame.

$$r^l = \xi^{-1}\left(\sum_{c=1}^{C} F^c \odot \overline{Z}^c\right) \tag{3.77}$$

Here, ξ^{-1} denotes the inverse FFT transform.

III One-Stream Paradigm Typical Method: HDETrack [195]
Existing event-camera-based tracking methods often attempt to enhance tracking performance through multi-modal data fusion or pure event-driven approaches. However, multi-modal methods [242, 243] require processing both RGB and event data at the same time. This greatly increases inference costs. As a result, their use is limited in scenarios that demand high real-time performance. On the other hand, pure event-based methods [245, 272] avoid the high computational cost of multi-modal approaches. However, they struggle with sparse event data in low-speed scenarios. This often results in unclear object contours. Additionally, issues such as noisy events and sparse spatial resolution can compromise their robustness.

To address these challenges, HDETrack introduces a cross-modal hierarchical knowledge distillation framework. It uses multi-modal and multi-view data for knowledge transfer. This process occurs during the training phase, as shown in Fig. 3.20. This enables the use of only event data during the testing phase to achieve efficient and low-latency tracking, while significantly improving robustness and accuracy.

Specifically, the teacher network takes RGB frames and event streams as inputs. It extracts template and search regions from both inputs. These regions are converted into feature embeddings using a projection layer. The embeddings are then fed into a unified Transformer backbone. This backbone performs feature extraction, interaction, and fusion simultaneously. Then, a tracking head uses these features to locate the target. In contrast, the student network only uses event data as input and applies hierarchical knowledge distillation to enhance its tracking performance:

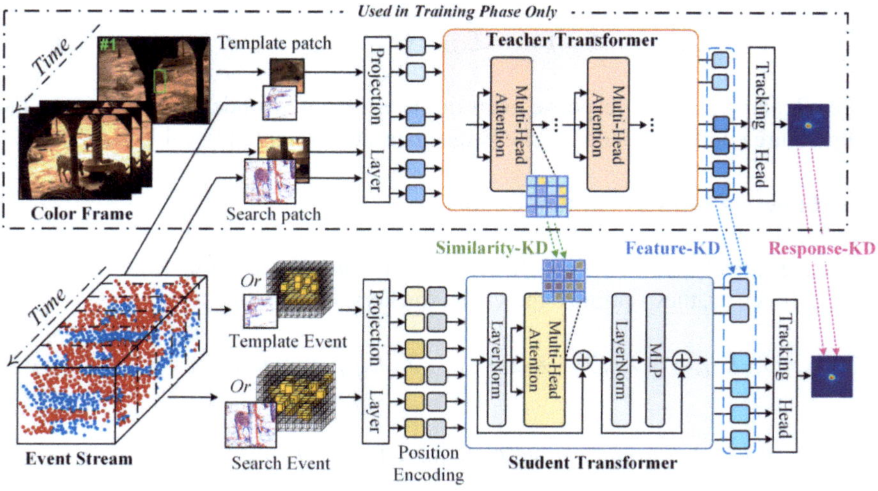

Fig. 3.20 The main architecture diagram of HDETrack [195]

Similarity Matrix Distillation Implementation Details
The similarity matrices computed by the teacher Transformer in its multi-head self-attention layers capture rich long-range and cross-modal relationships. By distilling these matrices from the teacher to the student network, the student can learn these critical relationships. The associated loss function is:

$$\mathcal{L}_{simKD} = \mathcal{L}_2\left(S_s^j, S_t^i\right) \tag{3.78}$$

Here, S_t^i is the similarity matrix of the ith layer in the teacher network, and S_s^j is the similarity matrix of the jth layer in the student network.

Feature Representation Distillation Implementation Details
A powerful feature representation is extracted from the teacher Transformer network and distilled into the student network to enhance the student's feature representation capability. The loss function for this process is:

$$\mathcal{L}_{featKD} = \|F_t - F_s\|_F^2 \tag{3.79}$$

Here, F_t is the feature representation of the teacher network, and F_s is the feature representation of the student network.

Response Map Distillation Implementation Details
The response maps output by the teacher network are used for target object localization. By imitating these response maps from the teacher network, the student network can better perform target localization. The loss function for this is:

$$\mathcal{L}_{resKD} = \mathcal{L}_{GWF}(R_s/\tau, R_t/\tau) \tag{3.80}$$

Here, R_t and R_s are the response maps of the teacher and student networks, τ is the temperature coefficient, and $\mathcal{L}_{GWF}$ is the Gaussian Weighted Focal loss function.

Loss Function Implementation Details
Finally, the loss function combines multiple tracking loss terms. These include $focal$, $\mathcal{L}_1$, and $GIoU$ losses. It also incorporates three knowledge distillation losses. Together, they optimize the visual tracker:

$$\mathcal{L}_{total} = \lambda_1\mathcal{L}_{focal} + \lambda_2\mathcal{L}_1 + \lambda_3\mathcal{L}_{GIoU} + \eta_1\mathcal{L}_{simKD} + \eta_2\mathcal{L}_{featKD} + \eta_3\mathcal{L}_{resKD} \tag{3.81}$$

Here, $\mathcal{L}_{focal}$ is the focal loss, $\mathcal{L}_1$ is the L1 loss, $\mathcal{L}_{GIoU}$ is the GIoU loss, and $\lambda_1, \lambda_2, \lambda_3$ along with η_1, η_2, η_3 are weight parameters that balance the different loss terms.

3.4.3 Event-Based VOT Methods Performance Comparison

Table 3.5 offers a performance comparison of several typical event-based VOT methods across different datasets. These methods belong to various paradigms. By evaluating metrics like accuracy and recall on datasets such as FE240hz and VisEvent, the performance differences among methods from different paradigms are clearly visible. Notably, the one-stream paradigm's HDETrack series methods show remarkable performance in accuracy and recall, highlighting their strong competitiveness in event-based VOT. Siamese network paradigm methods like DANet also display unique advantages. Overall, the table serves as a key reference for researchers in selecting and enhancing event-based VOT methods.

3.4.4 Summary and Challenges

This section mainly introduces single-modal VOT methods based on event modality. It covers the modality's characteristics, application advantages, and typical algorithms across different paradigms. Event cameras, inspired by biological vision systems, boast high temporal resolution, low latency, low power consumption, and a high dynamic range, excelling in dynamic and complex scenes. The section delves into four event-driven tracking paradigms: Siamese networks, correlation filters, one-stream, and detection tracking. It also focuses on some algorithms' principles, features, and strategies for handling complex scenes.

However, the application of VOT based on event modality still faces the following challenges:

1. **Modality Characteristic Difference Integration**: Event camera outputs are sparse event data differing structurally and in features from traditional RGB data. Effectively fusing event data with other modalities (e.g., RGB images, radar data) to enhance system performance by leveraging their complementary information remains a key challenge.
2. **Representation and Utilization of Sparse Data**: Event data are sparse in space and time. Designing effective feature extraction and representation methods is crucial. These methods should fully utilize data sparsity for target modeling and tracking. At the same time, they must prevent accuracy loss caused by the sparse input.
3. **Noise Interference in Dynamic Scenes**: In dynamic scenes, event cameras are prone to noise interference that can mislead tracking algorithms. Distinguishing target events from noise to improve tracking robustness is vital.
4. **Addressing Fast Motion and Appearance Changes**: Maintaining stable tracking during fast target motion or significant appearance changes (e.g., rotation, scaling, deformation) is challenging. Leveraging event data's fast dynamic change capture capability into effective tracking strategies requires further study.

Table 3.5 Comparison of the performance of different Event trackers

			FE240hz				VisEvent				
Paradigms	Trackers	Publication	SR	PR	$OP_{0.5}$	$OP_{0.75}$	SR	PR	NPR	$OP_{0.5}$	$OP_{0.75}$
Siamese Network	DANet[46]	TIP2023	56.86	89.19	69.94	25.03	39.76	54.54	–	45.62	30.27
One Stream	MambaEVT[196]	CVPR2024	58.09	91.97	–	–	35.9	50.2	39.4	–	–
	HDETrack[195]	CVPR2024	59.8	92.2	–	–	37.3	54.6	44.5	–	–
	HDETrack V2[194]	CVPR2025	60.4	93.2	–	–	38.0	55.8	45.1	–	–
Other	STNet[245]	CVPR2022	58.5	89.6	74.6	28.1	35.5	49.2	–	39.7	20.4

5. **Distinction and Localization in Multi-target Scenes**: Accurately identifying target objects in multi-target scenes is a key challenge. It is also important to avoid confusion between targets. Precise tracking becomes even harder during interaction or occlusion. These issues are critical for event modality tracking.
6. **Adaptability to Extreme Environments**: Event cameras have a high dynamic range but may be affected in extreme environments like low light. Optimizing algorithms to adapt to these conditions for stable and accurate tracking is necessary.
7. **Real-time and Efficiency Optimization of Algorithms**: For real-time applications like autonomous driving and robot navigation, speed is critical. Event modality tracking algorithms must improve both speed and efficiency. At the same time, they should maintain high accuracy. Reducing computational complexity and resource usage is also essential for real-time tracking.

References

1. Abdelaziz O (2024) Beyond traditional single object tracking. PhD thesis. University of British Columbia, Columbia
2. Alzugaray I, Chli M (2018) ACE: an efficient asynchronous corner tracker for event cameras. In: 2018 International Conference on 3D Vision (3DV). IEEE, New York, pp 653–661
3. Alzugaray I, Chli M (2020) Haste: multi-hypothesis asynchronous speeded-up tracking of events. In: 31st British Machine Vision Virtual Conference (BMVC 2020). ETH Zurich, Institute of Robotics and Intelligent Systems, Zurich, p 744
4. Amir A, et al (2017) A low power, fully event-based gesture recognition system. In: Proceedings of the IEEE conference on computer vision and pattern recognition, pp 7243–7252
5. CS Asha, Narasimhadhan AV (2017) Robust infrared target tracking using discriminative and generative approaches. Infrared Phys Technol 85:114–127
6. Aziz NH, et al (2023) Detection of bacterial leaf blight disease using RGB-based vegetation indices and fuzzy logic. In: 2023 19th IEEE international colloquium on signal processing & its applications (CSPA). IEEE, New York, pp 134–139
7. Barranco F, Fermuller C, Ros E (2018) Real-time clustering and multi-target tracking using event-based sensors. In: 2018 IEEE/RSJ international conference on intelligent robots and systems (IROS). IEEE, New York, pp 5764–5769
8. Baudron A, et al (2020) E3d: event-based 3d shape reconstruction. arXiv preprint arXiv:2012.05214
9. Behley J, et al (2019) Semantickitti: a dataset for semantic scene understanding of lidar sequences. In: Proceedings of the IEEE/CVF international conference on computer vision, pp 9297–9307
10. Bertinetto L, et al (2016) Fully-convolutional siamese networks for object tracking. In: Proceedings of the Computer vision–ECCV 2016 workshops: Amsterdam, the Nether-lands, October 8–10 and 15–16, 2016, part II 14. Springer, Berlin, pp 850–865
11. Bhat G, et al (2019) Learning discriminative model prediction for tracking. In: Proceedings of the IEEE/CVF international conference on computer vision, pp 6182–6191
12. Binas J, et al (2017) DDD17: End-to-end DAVIS driving dataset. arXiv preprint arXiv:1711.01458
13. Cai W, Liu Q, Wang Y (2024) Hiptrack: visual tracking with historical prompts. In: Proceedings of the IEEE/CVF Conference on Computer Vision and Pattern Recognition, pp 19258–19267

14. Cai W, Liu Q, Wang Y (2025) SPMTrack: spatio-temporal parameter-efficient fine-tuning with mixture of experts for scalable visual tracking. In: Proceedings of the computer vision and pattern recognition conference, pp 16871–16881
15. Cai Y, et al (2023) Robust object modeling for visual tracking. In: Proceedings of the IEEE/CVF international conference on computer vision, pp 9589–9600
16. Calabrese E, et al (2019) DHP19: dynamic vision sensor 3D human pose dataset. In: Proceedings of the IEEE/CVF conference on computer vision and pattern recognition workshops, pp 0–0
17. Cao Z, et al (2025) Siamese adaptive network-based accurate and robust visual object tracking algorithm for quadrupedal robots. In: IEEE Transactions on Cybernetics
18. Cazzato D, Bono F (2024) An application-driven survey on event-based neuromorphic computer vision. Information 15(8):472
19. Chae Y, Wang L, Yoon K-J (2021) Siamevent: event-based object tracking via edge-aware similarity learning with siamese networks. arXiv preprint arXiv:2109.13456
20. Chen B, et al (2022) Backbone is all your need: a simplified architecture for visual object tracking. In: European conference on computer vision. Springer, Berlin, pp 375–392
21. Chen H, et al (2019) Asynchronous tracking-by-detection on adaptive time surfaces for event-based object tracking. In: Proceedings of the 27th ACM International Conference on Multimedia, pp 473–481
22. Chen H, et al (2020) End-to-end learning of object motion estimation from retinal events for event-based object tracking. In: Proceedings of the AAAI conference on artificial intelligence, vol. 34(07), pp 10534–10541
23. Chen R, et al (2022) GFSNet: generalization-friendly siamese network for thermal infrared object tracking. Infrared Phys Technol123:104190
24. Chen S-F, et al (2025) Improving visual object tracking through visual prompting. IEEE Trans Multimedia 27:2682–2694
25. Chen X, et al (2023) Seqtrack: sequence to sequence learning for visual object tracking. In: Proceedings of the IEEE/CVF conference on computer vision and pattern recognition, pp 14572–14581
26. Chen X, et al (2021) Transformer tracking. In: Proceedings of the IEEE/CVF conference on computer vision and pattern recognition, pp 8126–8135
27. Chen Z, et al (2023) Segment any events via weighted adaptation of pivotal tokens. arXiv preprint arXiv:2312.16222
28. Cheng S, et al (2021) Learning to filter: siamese relation network for robust tracking. In: Proceedings of the IEEE/CVF conference on computer vision and pattern recognition, pp 4421–4431
29. Choi J, Kwon J, Lee KM (2019) Deep meta learning for real-time target-aware visual tracking. In: Proceedings of the IEEE/CVF international conference on computer vision, pp 911–920
30. Choi J, et al (2017) Attentional correlation filter network for adaptive visual tracking. In: Proceedings of the IEEE conference on computer vision and pattern recognition, pp 4807–4816
31. Cui Y, et al (2023) Mixformerv2: efficient fully transformer tracking. Adv Neural Inf Proces Syst 36:58736–58751
32. Dai J, et al (2025) An efficient object tracking based on multi-head cross-attention transformer. Expert Syst 42(2):e13650
33. Dai K, et al (2019) Visual tracking via adaptive spatially-regularized correlation filters. In: Proceedings of the IEEE/CVF conference on computer vision and pattern recognition, pp 4670–4679
34. Danelljan M, et al (2016) Beyond correlation filters: learning continuous convolution operators for visual tracking. In: Proceedings of the Computer Vision–ECCV 2016: 14th European conference, Amsterdam, The Netherlands, October 11–14, 2016, Part V 14. Springer, Berlin, pp 472–488

35. Danelljan M, et al (2015) Convolutional features for correlation filter based visual tracking. In: Proceedings of the IEEE international conference on computer vision workshops, pp 58–66
36. Danelljan M, et al (2017) Eco: Efficient convolution operators for tracking. In: Proceedings of the IEEE conference on computer vision and pattern recognition, pp 6638–6646
37. Dardelet L, Ieng S-H, Benosman R (2018) Event-based features selection and tracking from intertwined estimation of velocity and generative contours. arXiv preprint arXiv:1811.07839
38. Demir HS,Adil OF (2018) Part-based co-difference object tracking algorithm for infrared videos. In: 2018 25th IEEE International Conference on Image Processing (ICIP). IEEE, New York, pp 3723–3727
39. Deng J, et al (2022) Applying convolutional neural networks for detecting wheat stripe rust transmission centers under complex field conditions using RGB-based high spatial resolution images from UAVs. Comput Electron Agric 200:107211
40. Deng Y, Li Y, Chen H (2020) Amae: adaptive motion-agnostic encoder for event-based object classification. IEEE Rob Autom Lett 5(3):4596–4603
41. Deniz D, et al (2023) When do neuromorphic sensors outperform cameras? Learning from dynamic features. In: 2023 57th Annual Conference on Information Sciences and Systems (CISS). IEEE, New York, pp 1–6
42. Dong Y, et al (2023) Bullying10k: a large-scale neuromorphic dataset towards privacy-preserving bullying recognition. Adv Neural Inf Proces Syst 36, pp 1923–1937
43. Du F, et al (2020) Correlation-guided attention for corner detection based visual tracking. In: Proceedings of the IEEE/CVF conference on computer vision and pattern recognition, pp 6836–6845
44. Fan H, Ling H (2017) Parallel tracking and verifying: a framework for real-time and high accuracy visual tracking. In: Proceedings of the IEEE international conference on computer vision, pp 5486–5494
45. Fan H, Ling H (2019) Siamese cascaded region proposal networks for real-time visual tracking. In: Proceedings of the IEEE/CVF conference on computer vision and pattern recognition, pp 7952–7961
46. Fu Y, et al (2023) Distractor-aware event-based tracking. IEEE Trans Image Process 32:6129–6141
47. Fu Z, et al (2022) SparseTT: Visual tracking with sparse transformers. arXiv preprint arXiv:2205.03776
48. Fu Z, et al (2021) Stmtrack: template-free visual tracking with space-time memory networks. In: Proceedings of the IEEE/CVF conference on computer vision and pattern recognition, pp 13774–13783
49. Gallego G, et al (2020) Event-based vision: a survey. IEEE Trans Pattern Anal Mach Intell 44(1):154–180
50. Gao J, Zhang T, Xu C (2019) Graph convolutional tracking. In: Proceedings of the IEEE/CVF conference on computer vision and pattern recognition, pp 4649–4659
51. Gao S, Zhou C, Zhang J (2023) Generalized relation modeling for transformer tracking. In: Proceedings of the IEEE/CVF conference on computer vision and pattern recognition, pp 18686–18695
52. Gao Y, et al (2023) Action recognition and benchmark using event cameras. IEEE Trans Pattern Anal Mach Intell 45(12):14081–14097
53. Gao Y, et al (2024) Hypergraph-based multi-view action recognition using event cameras. IEEE Trans Pattern Anal Mach Intell 46(10):6610–6622
54. Gehrig D, et al (2019) End-to-end learning of representations for asynchronous event-based data. In: Proceedings of the IEEE/CVF international conference on computer vision, pp 5633–5643
55. Gerardo R, de Lima IP (2023) Applying RGB-based vegetation indices obtained from UAS imagery for monitoring the rice crop at the field scale: a case study in Portugal. Agriculture 13(10):1916

56. Giancola S, Zarzar J, Ghanem B (2019) Leveraging shape completion for 3d siamese tracking. In: Proceedings of the IEEE/CVF conference on computer vision and pattern recognition, pp 1359–1368
57. Glover A, Bartolozzi C (2017) Robust visual tracking with a freely-moving event camera. In: 2017 IEEE/RSJ international conference on intelligent robots and systems (IROS). IEEE, New York, pp 3769–3776
58. Glover A, et al (2021) luvharris: a practical corner detector for event-cameras. IEEE Trans Pattern Anal Mach Intell 44(12), 10087–10098
59. Gopal YG, Amer MA (2024) Separable self and mixed attention transformers for efficient object tracking. In: Proceedings of the IEEE/CVF winter conference on applications of computer vision, pp 6708–6717
60. Gu D, et al (2021) How to learn a domain-adaptive event simulator? In: Proceedings of the 29th ACM international conference on multimedia, pp 1275–1283
61. Gu D, et al (2023) Reliable event generation with invertible conditional normalizing flow. IEEE Trans Pattern Anal Mach Intell 46(2):927–943
62. Gundogdu E, et al (2015) Comparison of infrared and visible imagery for object tracking: toward trackers with superior ir performance. In: Proceedings of the IEEE conference on computer vision and pattern recognition workshops, pp 1–9
63. Gundogdu E, et al (2016) Evaluation of feature channels for correlation-filter-based visual object tracking in infrared spectrum. In: Proceedings of the IEEE conference on computer vision and pattern recognition workshops, pp 24–32
64. Guo D, et al (2021) Graph attention tracking. In: Proceedings of the IEEE/CVF conference on computer vision and pattern recognition, pp 9543–9552
65. Guo D, et al (2020) SiamCAR: siamese fully convolutional classification and regression for visual tracking. In: Proceedings of the IEEE/CVF conference on computer vision and pattern recognition, pp 6269–6277
66. Guo Q, et al (2017) Learning dynamic siamese network for visual object tracking. In: Proceedings of the IEEE international conference on computer vision, pp 1763–1771
67. Guo S, Delbruck T (2022) Low cost and latency event camera back-ground activity denoising. IEEE Trans Pattern Anal Mach Intell 45(1):785–795
68. Guo S, Gallego G (2024) CMax-SLAM: event-based rotational-motion bundle adjustment and SLAM system using contrast maximization. IEEE Trans Robot 40:2442–2461
69. Guo Y, et al (2020) Deep learning for 3d point clouds: a survey. IEEE Trans Pattern Anal Mach Intell 43(12):4338–4364
70. Gupta DK, Arya D, Gavves E (2021) Rotation equivariant siamese networks for tracking. In: Proceedings of the IEEE/CVF conference on computer vision and pattern recognition, pp 12362–12371
71. Han W, et al (2021) Learning to fuse asymmetric feature maps in siamese trackers. In: Proceedings of the IEEE/CVF conference on computer vision and pattern recognition, pp 16570–16580
72. He A, et al (2018) A twofold siamese network for real-time object tracking. In: Proceedings of the IEEE conference on computer vision and pattern recognition, pp 4834–4843
73. He K, et al (2016) Deep residual learning for image recognition. In: Proceedings of the IEEE conference on computer vision and pattern recognition, pp 770–778
74. He Y, et al (2015) Infrared target tracking based on robust low-rank sparse learning. IEEE Geosci Remote Sens Lett 13(2):232–236
75. He Y, et al (2021) Infrared machine vision and infrared thermography with deep learning: a review. Infrared Phys Technol 116:103754
76. He Z, et al (2017) Correlation filters with weighted convolution responses. In: Proceedings of the IEEE international conference on computer vision workshops, pp 1992–2000
77. Hong S, et al (2015) Online tracking by learning discriminative saliency map with convolutional neural network. In: International conference on machine learning. PMLR, New York, pp 597–606

78. Hu S, et al (2022) ECDT: event clustering for simultaneous feature detection and tracking. In: 2022 IEEE/RSJ international conference on intelligent robots and systems (IROS). IEEE, New York, pp 3808–3815
79. Hu Y, et al (2020) Ddd20 end-to-end event camera driving dataset: fusing frames and events with deep learning for improved steering prediction. In: 2020 IEEE 23rd international conference on intelligent transportation systems (ITSC). IEEE, New York, pp 1–6
80. Hu Y, et al (2016) DVS benchmark datasets for object tracking, action recognition, and object recognition. Front Neurosci 10:405
81. Huang C, Lucey S, Ramanan D (2017) Learning policies for adaptive tracking with deep feature cascades. In: Proceedings of the IEEE international conference on computer vision, pp 105–114
82. Huang K, et al (2023) Event-based simultaneous localization and mapping: a comprehensive survey. arXiv preprint arXiv:2304.09793
83. Huang L, Zhao X, Huang K (2019) Bridging the gap between detection and tracking: A unified approach. In: Proceedings of the IEEE/CVF international conference on computer vision, pp 3999–4009
84. Huang L, Zhao X, Huang K (2020) Globaltrack: a simple and strong baseline for long-term tracking. In: Proceedings of the AAAI conference on artificial intelligence, vol. 34(07), pp 11037–11044
85. Hui L, et al (2022) 3D siamese transformer network for single object tracking on point clouds. In: Proceedings of the Computer Vision–ECCV 2022: 17th European conference, Tel Aviv, Israel, October 23–27, 2022, Part II. Springer, Berlin, pp 293–310
86. Hui L, et al (2021) 3D Siamese voxel-to-BEV tracker for sparse point clouds. Adv Neural Inf Proces Syst 34:28714–28727
87. Ji M, et al (2023) SCTN: Event-based object tracking with energy-efficient deep convolutional spiking neural networks. Front Neurosci17:1123698
88. Jin L, Cheng J, Zhang C (2021) Infrared pedestrian tracking with graph memory features. IEEE Signal Process Lett 28:1933–1937
89. Joubert D, et al (2021) Event camera simulator improvements via characterized parameters. Front Neurosci15:702765
90. Kang B, et al (2023) Exploring lightweight hierarchical vision transformers for efficient visual tracking. In: Proceedings of the IEEE/CVF international conference on computer vision, pp 9612–9621
91. Kiani HG, Fagg A, Lucey S (2017) Learning background-aware correlation filters for visual tracking. In: Proceedings of the IEEE international conference on computer vision, pp 1135–1143
92. Kim C, Ko H (2020) Weighted kernel filter based anti-air object tracking for thermal infrared systems. Sensors 20(15):4081
93. Sachin Sakthi KS, Joo YH, Jeong JH (2025) Keypoint prediction enhanced Siamese networks with attention for accurate visual object tracking. Expert Syst Appl 268:126237
94. Lagorce X, et al (2014) Asynchronous event-based multikernel algorithm for high-speed visual features tracking. IEEE Trans Neural Networks Learn Syst 26(8):1710–1720
95. Lai S, et al (2024) Refocus the attention for parameter-efficient thermal infrared object tracking. IEEE Trans Neural Networks Learn Syst 36:9538–9549
96. Li B, et al (2018) High performance visual tracking with siamese region proposal network. In: Proceedings of the IEEE conference on computer vision and pattern recognition, pp 8971–8980
97. Li B, et al (2019) Siamrpn++: evolution of siamese visual tracking with very deep networks. In: Proceedings of the IEEE/CVF conference on computer vision and pattern recognition, pp 4282–4291
98. Li, et al (2017) Integrating boundary and center correlation filters for visual tracking with aspect ratio variation. In: Proceedings of the IEEE international conference on computer vision workshops, pp 2001–2009

99. Li F, et al (2018) Learning spatial-temporal regularized correlation filters for visual tracking. In: Proceedings of the IEEE conference on computer vision and pattern recognition, pp 4904–4913
100. Li H, Shi L (2019) Robust event-based object tracking combining correlation filter and CNN representation. Front Neurorobot 13:82
101. Li J, et al (2025) Robust underwater object tracking with image enhancement and two-step feature compression. Complex Intell Syst 11(2):1–14
102. Li M, et al (2019) Mask sparse representation based on semantic features for thermal infrared target tracking. Remote Sens 11(17):1967
103. Li P, et al (2019) GradNet: gradient-guided network for visual object tracking. In: Proceedings of the IEEE/CVF International conference on computer vision, pp 6162–6171
104. Li Q, et al (2024) Progressive domain adaptation for thermal infrared object tracking. arXiv preprint arXiv:2407.19430
105. Li R, et al (2019) Fa-harris: a fast and asynchronous corner detector for event cameras. In: 2019 IEEE/RSJ international conference on intelligent robots and systems (IROS). IEEE, New York, pp 6223–6229
106. Li S, et al (2024) Two-stage spatio-temporal feature correlation network for infrared ground target tracking. IEEE Trans Geosci Remote Sens 62:1–14
107. Li S, et al (2020) TLPG-Tracker: joint learning of target localization and proposal generation for visual tracking. In: *IJCAI*, pp 708–715
108. Li X, et al (2019) Hierarchical spatial-aware siamese network for thermal infrared object tracking. Knowl-Based Syst 166:71–81
109. Li X, et al (2019) Target-aware deep tracking. In: Proceedings of the IEEE/CVF conference on computer vision and pattern recognition, pp 1369–1378
110. Li Y, et al (2020) AutoTrack: towards high-performance visual tracking for UAV with automatic spatio-temporal regularization. In: Proceedings of the IEEE/CVF conference on computer vision and pattern recognition, pp 11923–11932
111. Li YC, Yang S (2022) Infrared small object tracking based on Att-Siam network. IEEE Access 10:133766–133777
112. Li Z, et al (2023) Motion-to-matching: a mixed paradigm for 3D single object tracking. IEEE Rob Autom Lett 9(2):1468–1475
113. Li Z, et al (2023) Motion-to-matching: a mixed paradigm for 3d single object tracking. IEEE Rob Autom Lett 9(2):1468–1475
114. Liang S, et al (2025) Autoregressive sequential pretraining for visual tracking. In: Proceedings of the computer vision and pattern recognition conference (CVPR), pp 7254–7264
115. Liao B, et al (2020) Pg-net: Pixel to global matching network for visual tracking. In: Proceedings of the computer vision–ECCV 2020: 16th European Conference, Glasgow, UK, August 23–28, 2020, Part XXII 16. Springer, Berlin, pp 429–444
116. Liao D, et al (2025) Fine-grained feature and template reconstruction for TIR object tracking. In: IEEE transactions on circuits and systems for video technology
117. Lin S, et al (2022) Dvs-voltmeter: stochastic process-based event simulator for dynamic vision sensors. In: European Conference on Computer Vision. Springer, Berlin, pp 578–593
118. Litzenberger M, et al (2006) Embedded vision system for real-time object tracking using an asynchronous transient vision sensor. In: 2006 IEEE 12th digital signal processing workshop & 4th IEEE signal processing education workshop. IEEE, New York, pp 173–178
119. Liu C, et al (2022) Fast classification and action recognition with event-based imaging. IEEE Access 10:55638–55649
120. Liu J, et al (2024) M3SOT: multi-frame, multi-field, multi-space 3D single object tracking. In: Proceedings of the AAAI conference on artificial intelligence, vol. 38(4), pp 3630–3638
121. Liu M, Delbruck T (2022) EDFLOW: event driven optical flow camera with keypoint detection and adaptive block matching. IEEE Trans Circuits Syst Video Technol 32(9):5776–5789
122. Liu N, Zhang Y (2025) A transformer based visual tracker with restricted token interaction and knowledge distillation. Knowl-Based Syst 307:112736

123. Liu Q, et al (2017) Deep convolutional neural networks for thermal infrared object tracking. Knowl-Based Syst 134:189–198
124. Liu Q, et al (2020) Learning deep multi-level similarity for thermal infrared object tracking. IEEE Trans Multimedia 23:2114–2126
125. Liu Q, et al (2022) Learning dual-level deep representation for thermal infrared tracking. IEEE Trans Multimedia 25:1269–1281
126. Liu R, Lu Y (2012) Infrared target tracking in multiple feature pseudo-color image with kernel density estimation. Infrared Phys Technol 55(6):505–512
127. Liu Y, Jiang J, Sun J (2021) Hand pose estimation from rgb images based on deep learning: A survey. In: 2021 IEEE 7th international conference on virtual reality (ICVR). IEEE, New York, pp 82–89
128. Lu X, et al (2018) Deep regression tracking with shrinkage loss. In: Proceedings of the European conference on computer vision (ECCV), pp 353–369
129. Lukezic A, et al (2017) Discriminative correlation filter with channel and spatial reliability. In: Proceedings of the IEEE conference on computer vision and pattern recognition, pp 6309–6318
130. Lungu I-A, Corradi F, Delbrück T (2017) Live demonstration: convolutional neural network driven by dynamic vision sensor playing RoShamBo. In: 2017 IEEE international symposium on circuits and systems (ISCAS). IEEE, New York, pp 1–1
131. Luo Z, et al (2024) Modeling continuous motion for 3d point cloud object tracking. In: Proceedings of the AAAI conference on artificial intelligence, vol. 38(5), pp 4026–4034
132. Ma C, et al (2015) Hierarchical convolutional features for visual tracking. In: Proceedings of the IEEE international conference on computer vision, pp 3074–3082
133. Ma D, Wu X (2021) CapsuleRRT: relationships-aware regression tracking via capsules. In: Proceedings of the IEEE/CVF conference on computer vision and pattern recognition, pp 10948–10957
134. Ma S, et al (2025) AMTrack: transformer tracking via action information and mix-frequency features. Expert Syst Appl 261:125451
135. Ma T, et al (2023) Synchronize feature extracting and matching: a single branch framework for 3d object tracking. In: Proceedings of the IEEE/CVF international conference on computer vision, pp 9953–9963
136. Manderscheid J, et al (2019) Speed invariant time surface for learning to detect corner points with event-based cameras. In: Proceedings of the IEEE/CVF conference on computer vision and pattern recognition, pp 10245–10254
137. Mayer C, et al (2022) Transforming model prediction for tracking. In: Proceedings of the IEEE/CVF conference on computer vision and pattern recognition, pp 8731–8740
138. Meshgi K, Oba S, Ishii S (2018) Efficient diverse ensemble for discriminative co-tracking. In: Proceedings of the IEEE conference on computer vision and pattern recognition, pp 4814–4823
139. Micev K, et al (2024) Measuring diameters and velocities of artificial rain-drops with a neuromorphic event camera. Atmos Meas Tech 17(1):335–357
140. Moeys PD, et al (2016) Steering a predator robot using a mixed frame/event-driven convolutional neural network. In: 2016 Second international conference on event-based control, communication, and signal processing (EBCCSP). IEEE, New York, pp 1–8
141. Mohamed SAS, et al (2021) Dynamic resource-aware corner detection for bio-inspired vision sensors. In: 2020 25th International conference on pattern recognition (ICPR). IEEE, New York, pp 10465–10472
142. Mueller M, Smith N, Ghanem B (2017) Context-aware correlation filter tracking. In: Proceedings of the IEEE conference on computer vision and pattern recognition, pp 1396–1404
143. Ni Z, et al (2015) Visual tracking using neuromorphic asynchronous event-based cameras. Neural Comput 27(4):925–953

144. Nie J, et al (2023) GLT-T: global-local transformer voting for 3D single object tracking in point clouds. In: Proceedings of the AAAI Conference on Artificial Intelligence, pp 1957–1965
145. Nie J, et al (2023) OSP2B: one-stage point-to-box network for 3D siamese tracking. In: Proceedings of the thirty-second international joint conference on artificial intelligence, pp 1285–1293
146. Nie J, et al (2024) P2P: part-to-part motion cues guide a strong tracking framework for LiDAR point clouds. arXiv preprint arXiv:2407.05238
147. Nie J, et al (2024) Towards category unification of 3d single object tracking on point clouds. arXiv preprint arXiv:2401.11204
148. Ning X, et al (2024) RGB-based set prediction transformer of 6D pose estimation for robotic grasping application. In: IEEE Access 12:138047–138060
149. Pan L, et al (2019) Bringing a blurry frame alive at high frame-rate with an event camera. In: Proceedings of the IEEE/CVF conference on computer vision and pattern recognition, pp 6820–6829
150. Pu S, et al (2018) Deep attentive tracking via reciprocative learning. Adv Neural Inf Proces Syst 31:1931–1941
151. Qi CR, et al (2017) Pointnet: deep learning on point sets for 3d classification and segmentation. In: Proceedings of the IEEE conference on computer vision and pattern recognition, pp 652–660
152. Qi CR, et al (2017) Pointnet + +: Deep hierarchical feature learning on point sets in a metric space. Adv Neural Inf Proces Syst 30:5099–5108
153. Qi H, et al (2020) P2b: Point-to-box network for 3d object tracking in point clouds. In: Proceedings of the IEEE/CVF conference on computer vision and pattern recognition, pp 6329–6338.
154. Qi M, et al (2024) Exploring reliable infrared object tracking with spatio-temporal fusion transformer. Knowl-Based Syst 284:111234
155. Qi Y, et al (2016) Hedged deep tracking. In: Proceedings of the IEEE conference on computer vision and pattern recognition, pp 4303–4311
156. Qi Y, et al (2024) Deblurring neural radiance fields with event-driven bundle adjustment. In: Proceedings of the 32nd ACM international conference on multimedia, pp 9262–9270
157. Ramesh B, et al (2020) e-tld: Event-based framework for dynamic object tracking. IEEE Trans Circuits Syst Video Technol 31(10):3996–4006
158. Rebecq H, Gehrig D, Scaramuzza D (2018) ESIM: an open event camera simulator. In: Conference on robot learning. PMLR, New York, pp 969–982
159. Ren W, et al (2023) Visual semantic segmentation based on few/zero-shot learning: an overview. IEEE/CAA J Autom Sin 11(5):1106–1126
160. Reza MN, et al (2024) RGB-based machine vision for enhanced pig disease symptoms monitoring and health management: a review. J Anim Sci Technol 67(1):17
161. Rueckauer B, Delbruck T (2016) Evaluation of event-based algorithms for optical flow with ground-truth from inertial measurement sensor. Front Neurosci 10:176
162. Rusu RB, Cousins S (2011) 3d is here: point cloud library (PCL). In: 2011 IEEE international conference on robotics and automation. IEEE, New York, pp 1–4
163. Shan Y, et al (2025) SDTrack: A baseline for event-based tracking via spiking neural networks. arXiv preprint arXiv:2503.08703
164. Shi L, et al (2024) Explicit visual prompts for visual object tracking. In: Proceedings of the AAAI conference on artificial intelligence, vol. 38(5), pp 4838–4846
165. Simonyan K, Zisserman A (2014) Very deep convolutional networks for large-scale image recognition. arXiv preprint arXiv:1409.1556
166. Singh B, et al (2024) Non-destructive chlorophyll prediction by machine learning techniques using RGB-based vegetation indices in wheat. In: Plant Physiology Reports , pp 1–12
167. Song J, et al (2023) Infrared object-tracking algorithm with temporal features fusion Siamese network. IEEE Trans Instrum Meas 73:1–13

168. Song Y, et al (2017) Crest: convolutional residual learning for visual tracking. In: Proceedings of the IEEE international conference on computer vision, pp 2555–2564
169. Song Y, et al (2018) Vital: visual tracking via adversarial learning. In: Proceedings of the IEEE conference on computer vision and pattern recognition, pp 8990–8999
170. Song Z, et al (2022) Transformer tracking with cyclic shifting window attention. In: Proceedings of the IEEE/CVF conference on computer vision and pattern recognition, pp 8791–8800
171. Sun J, et al (2021) Unsupervised cross-modal distillation for thermal infrared tracking. In: Proceedings of the 29th ACM international conference on multimedia, pp 2262–2270
172. Sun L, et al (2025) Visual object tracking based on adaptive deblurring integrating motion blur perception. In: Journal of visual communication and image representation , p 104388
173. Sun P-SV, et al (2024) Memory efficient corner detection for event-driven dynamic vision sensors. In: ICASSP 2024-2024 IEEE international conference on acoustics, speech and signal processing (ICASSP). IEEE, New York, pp 13416–13420
174. Sun X, et al (2017) Non-rigid object tracking via deformable patches using shape-preserved KCF and level sets. In: Proceedings of the IEEE international conference on vision C, pp 5495–5503
175. Sun Y, et al (2019) Roi pooled correlation filters for visual tracking. In: Proceedings of the IEEE/CVF conference on computer vision and pattern recognition, pp 5783–5791
176. Supancic J III and Ramanan D (2017) Tracking as online decision-making: learning a policy from streaming videos with reinforcement learning. In: Proceedings of the IEEE international conference on computer vision, pp 322–331
177. Tang M, et al (2018) High-speed tracking with multi-kernel correlation filters. In: Proceedings of the IEEE conference on computer vision and pattern recognition, pp 4874–4883
178. Tao R, Gavves E, Smeulders AWM (2016) Siamese instance search for tracking. In: Proceedings of the IEEE conference on computer vision and pattern recognition, pp 1420–1429
179. Thomas H, et al (2019) Kpconv: flexible and deformable convolution for point clouds. In: Proceedings of the IEEE/CVF international conference on computer vision, pp 6411–6420
180. Trevisan LR, et al (2023) Estimating black oat biomass using digital surface models and a vegetation index derived from RGB-based aerial images. Remote Sens 15(5):1363
181. Valmadre J, et al (2017) End-to-end representation learning for correlation filter based tracking. In: Proceedings of the IEEE conference on computer vision and pattern recognition, pp 2805–2813
182. Voigtlaender P, et al (2020) Siam R-CNN: visual tracking by re-detection. In: Proceedings of the IEEE/CVF conference on computer vision and pattern recognition, pp 6578–6588
183. Vrekoussis T, et al (2009) Image analysis of breast cancer immunohistochemistry-stained sections using ImageJ: an RGB-based model. Anticancer Res 29(12):4995–4998
184. Wang G, et al (2019) Spm-tracker: series-parallel matching for real-time visual object tracking. In: Proceedings of the IEEE/CVF conference on computer vision and pattern recognition, pp 3643–3652
185. Wang G, et al (2020) Tracking by instance detection: a meta-learning approach. In: Proceedings of the IEEE/CVF conference on computer vision and pattern recognition, pp 6288–6297
186. Wang J, et al (2025) OmniTracker: unifying visual object tracking by tracking-with-detection. IEEE Trans Pattern Anal Mach Intell 47(4):3159–3174. https://doi.org/10.1109/TPAMI.2025.3529926
187. Wang J, et al (2025) OmniTracker: unifying visual object tracking by tracking-with-detection. In: IEEE transactions on pattern analysis and machine intelligence
188. Wang M, Liu Y, Huang Z (2017) Large margin object tracking with circulant feature maps. In: Proceedings of the IEEE conference on computer vision and pattern recognition, pp 4021–4029
189. Wang M, et al (2023) Correlation pyramid network for 3d single object tracking. In: Proceedings of the IEEE/CVF conference on computer vision and pattern recognition, pp 3216–3225

190. Wang N, et al (2018) Multi-cue correlation filters for robust visual tracking. In: Proceedings of the IEEE conference on computer vision and pattern recognition, pp 4844–4853
191. Wang Q, et al (2019) Fast online object tracking and segmentation: A unifying approach. In: Proceedings of the IEEE/CVF conference on computer vision and pattern recognition, pp 1328–1338
192. Wang Q, et al (2018) Learning attentions: residual attentional Siamese network for high performance online visual tracking. In: Proceedings of the IEEE conference on computer vision and pattern recognition, pp 4854–4863
193. Wang Q, Yang P, Dou L (2025) Enhancing visual tracking with multi-scale cross-correlation and progressive attention. Neurocomputing 623:129417
194. Wang S, et al (2025) Event stream-based visual object tracking: HDETrack V2 and a high-definition benchmark. arXiv preprint arXiv:2502.05574
195. Wang X, et al (2024) Event stream-based visual object tracking: a high-resolution benchmark dataset and a novel baseline. In: Proceedings of the IEEE/CVF conference on computer vision and pattern recognition, pp 19248–19257
196. Wang X, et al (2024) Mambaevt: event stream based visual object tracking using state space model. arXiv preprint arXiv:2408.10487
197. Wang X, et al (2018) Sint + +: robust visual tracking via adversarial positive instance generation. In: Proceedings of the IEEE conference on computer vision and pattern recognition, pp 4864–4873
198. Wang X, Wang H, Tian C (2021) Infrared image brightness correction for TIR object tracking. In: 2021 7th International conference on computer and communications (ICCC). IEEE, New York, pp 704–708
199. Wang Y, et al (2021) Thermal infrared object tracking based on adaptive feature fusion. In: 2021 11th International conference on information technology in medicine and education (ITME). IEEE, New York, pp 71–75
200. Wang Z, et al (2025) ADSTrack: adaptive dynamic sampling for visual tracking. Complex Intell Syst 11(1):79
201. Wang Z, et al (2021) Mlvsnet: multi-level voting siamese network for 3d visual tracking. In: Proceedings of the IEEE/CVF international conference on computer vision, pp 3101–3110
202. Wei X, et al (2023) Autoregressive visual tracking. In: Proceedings of the IEEE/CVF conference on computer vision and pattern recognition, pp 9697–9706
203. Woo S, et al (2018) Cbam: convolutional block attention module. In: Proceedings of the European conference on computer vision (ECCV), pp 3–19
204. Wu Q, et al (2023) Dropmae: masked autoencoders with spatial-attention dropout for tracking tasks. In: Proceedings of the IEEE/CVF conference on computer vision and pattern recognition, pp 14561–14571
205. Wu S, et al (2021) Aircraft tracking in infrared imagery with adaptive learning and interference suppression. Electron Lett 57(16):636–638
206. Wu S, et al (2021) Joint feature embedding learning and correlation filters for aircraft tracking with infrared imagery. Neurocomputing 450:104–118
207. Xie F, Wang Z, Ma C (2024) Diffusiontrack: point set diffusion model for visual object tracking. In: Proceedings of the IEEE/CVF conference on computer vision and pattern recognition, pp 19113–19124
208. Xie F, et al (2022) Correlation-aware deep tracking. In: Proceedings of the IEEE/CVF conference on computer vision and pattern recognition, pp 8751–8760
209. Xie F, et al (2023) Videotrack: learning to track objects via video transformer. In: Proceedings of the IEEE/CVF conference on computer vision and pattern recognition, pp 22826–22835
210. Xie J, et al (2024) Autoregressive queries for adaptive tracking with spatio-temporal transformers. In: Proceedings of the IEEE/CVF conference on computer vision and pattern recognition, pp 19300–19309
211. Xu C, et al (2025) Less is more: token context-aware learning for object tracking. arXiv preprint arXiv:2501.00758

212. Xu T-X, et al (2023) CXTrack: improving 3D point cloud tracking with contextual information. In: Proceedings of the IEEE/CVF conference on computer vision and pattern recognition, pp 1084–1093
213. Xu T-X, et al (2023) Mbptrack: improving 3d point cloud tracking with memory networks and box priors. In: Proceedings of the IEEE/CVF international conference on computer vision, pp 9911–9920
214. Xu T, et al (2019) Joint group feature selection and discriminative filter learning for robust visual object tracking. In: Proceedings of the IEEE/CVF international conference on computer vision, pp 7950–7960
215. Xu Y, et al (2020) SiamFC + +: towards robust and accurate visual tracking with target estimation guidelines. In: Proceedings of the AAAI conference on artificial intelligence, vol. 34(07), pp 12549–12556
216. Xu Y, et al (2021) Hierarchical convolution fusion-based adaptive Siamese network for infrared target tracking. IEEE Trans Instrum Meas 70:1–12
217. Yan B, et al (2019) 'skimming-perusal'tracking: a framework for real-time and robust long-term tracking. In: Proceedings of the IEEE/CVF international conference on computer vision, pp 2385–2393
218. Yan B, et al (2021) Alpha-refine: boosting tracking performance by precise bounding box estimation. In: Proceedings of the IEEE/CVF conference on computer vision and pattern recognition, pp 5289–5298
219. Yang C-Y, et al (2024) Samurai: adapting segment anything model for zero-shot visual tracking with motion-aware memory. arXiv preprint arXiv:2411.11922
220. Yang D, et al (2023) Foreground-background distribution modeling transformer for visual object tracking. In: Proceedings of the IEEE/CVF international conference on computer vision, pp 10117–10127
221. Yang T, Chan AB (2017) Recurrent filter learning for visual tracking. In: Proceedings of the IEEE international conference on computer vision workshops, pp 2010–2019
222. Yang X, et al (2025) Adaptively bypassing vision transformer blocks for efficient visual tracking. Pattern Recogn 161:111278
223. Yao T, et al (2021) Scale and appearance variation enhanced siamese network for thermal infrared target tracking. Infrared Phys Technol 117:103825
224. Yao Y, et al (2018) Joint representation and truncated inference learning for correlation filter based tracking. In: Proceedings of the European conference on computer vision (ECCV), pp 552–567
225. Ye B, et al (2022) Joint feature learning and relation modeling for tracking: a one-stream framework. In: European conference on computer vision. Springer, Berlin, pp 341–357
226. Yeo D, et al (2017) Superpixel-based tracking-by-segmentation using markov chains. In: Proceedings of the IEEE conference on computer vision and pattern recognition, pp 1812–1821
227. Yu T, et al (2019) Robust thermal infrared object tracking with continuous correlation filters and adaptive feature fusion. Infrared Phys Technol 98:69–81
228. Yu W, et al (2023) Metaformer baselines for vision. IEEE Trans Pattern Anal Mach Intell 46(2):896–912
229. Yu W, et al (2022) Metaformer is actually what you need for vision. In: Proceedings of the IEEE/CVF conference on computer vision and pattern recognition, pp 10819–10829
230. Yu X, et al (2017) Dense structural learning for infrared object tracking at 200+ frames per second. Pattern Recogn Lett 100:152–159
231. Yu Y, et al (2020) Deformable siamese attention networks for visual object tracking. In: Proceedings of the IEEE/CVF conference on computer vision and pattern recognition, pp 6728–6737
232. Yuan D, et al (2024) Hierarchical attention siamese network for thermal infrared target tracking. IEEE Trans Instrum Meas 73:1–11
233. Yuan D, et al (2022) Structural target-aware model for thermal infrared tracking. Neurocomputing 491:44–56

234. Yuan D, et al (2023) Thermal infrared target tracking: a comprehensive review. IEEE Trans Instrum Meas 73:1–19
235. Yue R, Tian Z, Du S (2022) Action recognition based on RGB and skeleton data sets: a survey. Neurocomputing 512:287–306
236. Yun S, et al (2017) Action-decision networks for visual tracking with deep reinforcement learning. In: Proceedings of the IEEE conference on computer vision and pattern recognition, pp 2711–2720
237. Zaveri R, et al (2024) Improving accuracy and generalization for efficient visual tracking. arXiv preprint arXiv:2411.18855
238. Zhang D, et al (2021) Visual tracking via hierarchical deep reinforcement learning. In: Proceedings of the AAAI conference on artificial intelligence, vol 35(4), pp 3315–3323
239. Zhang H, Yin Z, Zhang H (2023) Thermal infrared object tracking using correlation filters improved by level set. Signal, Image Video Process 17(3):791–797
240. Zhang H, et al (2025) AwareTrack: object awareness for visual tracking via templates interaction. Image Vis Comput 154:105363
241. Zhang H, et al (2025) SAM-assisted temporal-location enhanced transformer segmentation for object tracking with online motion inference. Neurocomputing 617:128914
242. Zhang J, et al (2023) Frame-event alignment and fusion network for high frame rate tracking. In: Proceedings of the IEEE/CVF conference on computer vision and pattern recognition, pp 9781–9790
243. Zhang J, et al (2021) Object tracking by jointly exploiting frame and event domain. In: Proceedings of the IEEE/CVF international conference on computer vision, pp 13043–13052
244. Zhang J, et al (2025) Spiking neural networks with adaptive membrane time constant for event-based tracking. IEEE Trans Image Process 34:1009–1021
245. Zhang J, et al (2022) Spiking transformers for event-based single object tracking. In: Proceedings of the IEEE/CVF conference on Computer Vision and Pattern Recognition, pp 8801–8810
246. Zhang L, et al (2019) Learning the model update for siamese trackers. In: Proceedings of the IEEE/CVF international conference on computer vision, pp 4010–4019
247. Zhang L, et al (2018) Synthetic data generation for end-to-end thermal infrared tracking. IEEE Trans Image Process 28(4):1837–1850
248. Zhang M, et al (2018) Visual tracking via spatially aligned correlation filters network. In: Proceedings of the European conference on computer vision (ECCV), pp 469–485
249. Zhang S, et al (2022) Evtracker: an event-driven spatiotemporal method for dynamic object tracking. Sensors 22(16):6090
250. Zhang T, Xu C, Yang M-H (2017) Multi-task correlation particle filter for robust object tracking. In: Proceedings of the IEEE conference on computer vision and pattern recognition, pp 4335–4343
251. Zhang W, Song R, Li Y, et al (2020) Online decision based visual tracking via reinforcement learning. Adv Neural Inf Proces Syst 33:11778–11788
252. Zhang Y, et al (2018) Structured siamese network for real-time visual tracking. In: Proceedings of the European conference on computer vision (ECCV), pp 351–366
253. Zhang Z, Peng H (2019) Deeper and wider siamese networks for real-time visual tracking. In: Proceedings of the IEEE/CVF conference on computer vision and pattern recognition, pp 4591–4600
254. Zhang Z, et al (2021) Learn to match: automatic matching network design for visual tracking. In: Proceedings of the IEEE/CVF international conference on computer vision, pp 13339–13348
255. Zhang Z, et al (2020) Ocean: object-aware anchor-free tracking. In: European conference on computer vision. Springer, Berlin, pp 771–787
256. Zhang Z, et al (2024) V2ce: video to continuous events simulator. In: 2024 IEEE international conference on robotics and automation (ICRA). IEEE, New York, pp 12455–12461
257. Zhao L, et al (2025) Global–local feature-mixed network with template update for visual tracking. Pattern Recogn Lett 188:111–116

258. Zheng C, et al (2022) Beyond 3d siamese tracking: A motion-centric paradigm for 3d single object tracking in point clouds. In: Proceedings of the IEEE/CVF conference on computer vision and pattern recognition, pp 8111–8120
259. Zheng C, et al (2021) Box-aware feature enhancement for single object tracking on point clouds. In: Proceedings of the IEEE/CVF international conference on computer vision, pp 13199–13208
260. Zheng L, et al (2020) Thermal infrared pedestrian tracking using joint siamese network and exemplar prediction model. Pattern Recogn Lett 140:66–72
261. Zheng Y, et al (2024) Odtrack: online dense temporal token learning for visual tracking. In: Proceedings of the AAAI conference on artificial intelligence, vol. 38(7), pp 7588–7596
262. Zheng Z, He X, Weng J (2015) Approaching camera-based real-world navigation using object recognition. Procedia Comput Sci 53:428–436
263. Zhou J, Wang P, Sun H (2020) Discriminative and robust online learning for siamese visual tracking. In: Proceedings of the AAAI conference on artificial intelligence, vol. 34(07), pp 13017–13024
264. Zhou L, et al (2021) Model uncertainty guides visual object tracking. In: Proceedings of the AAAI conference on artificial intelligence, vol. 35(4), pp 3581–3589
265. Zhou X, et al (2023) Reading relevant feature from global representation memory for visual object tracking. Adv Neural Inf Proces Syst 36:10814–10827
266. Zhou Y, Tuzel O (2018) Voxelnet: end-to-end learning for point cloud based 3d object detection. In: Proceedings of the IEEE conference on computer vision and pattern recognition, pp 4490–4499
267. Zhou Z, et al (2021) Saliency-associated object tracking. In: Proceedings of the IEEE/CVF international conference on computer vision, pp 9866–9875
268. Zhu ZA, et al (2019) Unsupervised event-based learning of optical flow, depth, and egomotion. In: Proceedings of the IEEE/CVF conference on computer vision and pattern recognition, pp 989–997
269. Zhu Z, et al (2018) Distractor-aware siamese networks for visual object tracking. In: Proceedings of the European conference on computer vision (ECCV), pp 101–117
270. Zhu Z, et al (2018) End-to-end flow correlation tracking with spatial-temporal attention. In: Proceedings of the IEEE conference on computer vision and pattern recognition, pp 548–557
271. Zhu Z, et al (2017) UCT: learning unified convolutional networks for real-time visual tracking. In: Proceedings of the IEEE international conference on computer vision workshops, pp 1973–1982
272. Zhu Z, Hou J, Lyu X (2022) Learning graph-embedded key-event back-tracing for object tracking in event clouds. Adv Neural Inf Proces Syst 35:7462–7476
273. Zitnick CL, Dollár P (2014) Edge boxes: locating object proposals from edges. In: Proceedings of the computer Vision–ECCV 2014: 13th European Conference, Zurich, Switzerland, September 6–12, 2014, Part V 13. Springer, Berlin, pp 391–405
274. Zuo Y-F, et al (2024) Cross-modal Semidense 6-DOF tracking of an event camera in challenging conditions. IEEE Trans Robot 40:1600–1616

Chapter 4
Multi-Modal Object Tracking

Abstract This chapter focuses on multi-modal visual object tracking (VOT), exploring the integration of different data modalities such as RGB, depth, thermal, LiDAR, and language to enhance tracking robustness in complex environments. Multi-modal VOT is vital for intelligent systems operating under challenging conditions, including low light and occlusions. The chapter introduces key fusion strategies—early, mid-level, and late fusion—detailing their mechanisms, benefits, and trade-offs. It systematically categorizes tracking methods by modality pairings and fusion stages, analyzing their applications in indoor, outdoor, and cross-modal scenarios. Experimental comparisons are presented to illustrate performance differences. Additionally, the chapter identifies major challenges in the field, such as high computational costs, feature misalignment, limited data availability, and poor cross-domain transferability. The chapter concludes by outlining future research directions aimed at improving efficiency, adaptability, and generalization. Through comprehensive analysis and structured exposition, this chapter offers essential insights for advancing robust and versatile multi-modal tracking systems.

4.1 Multi-Modal VOT Overview

4.1.1 RGB-Depth Tracking in VOT

The RGB-Depth combination uses color from RGB cameras and depth data from sensors like Kinect to show both appearance and spatial structure. This synergy improves tracking precision in indoor settings by mitigating occlusions and scale variations, as depth data complement RGB's sensitivity to lighting changes. The resulting 3D spatial awareness allows for accurate localization. This makes it ideal for applications such as human–robot interaction, augmented reality, and indoor surveillance. In these cases, precise depth estimation improves target discrimination.

M. Wang et al., *Visual Object Tracking across Modalities*, Advances in Computer Vision and Pattern Recognition, https://doi.org/10.1007/978-981-95-3664-1_4

4.1.2 RGB-Language Tracking in VOT

The RGB-Language combination uses textual or linguistic input, like object descriptions or commands, along with RGB images. This allows for semantic-guided tracking. The method uses language to improve target identification. It helps trackers focus on specific objects based on context, even in busy scenes. This approach is used in human–computer interaction, smart homes, and assistive robotics. In these areas, verbal instructions help track specific items, making the system more responsive to user intent.

4.1.3 RGB-LiDAR Tracking in VOT

RGB-LiDAR pairing integrates high-resolution color imagery with the sparse, high-precision 3D point clouds from LiDAR, offering a powerful solution for outdoor tracking. LiDAR's ability to penetrate adverse weather and provide accurate distance measurements compensates for RGB's vulnerability to illumination and textureless surfaces. This combination works well in autonomous driving and urban navigation. It allows for reliable tracking of vehicles and pedestrians in different conditions. It uses LiDAR's detailed geometric data and the rich context from RGB images.

4.1.4 RGB-Thermal Tracking in VOT

The RGB-Thermal modality combines visible light data with thermal imaging. Thermal imaging works well in low-light or nighttime conditions where RGB cameras struggle. It detects heat signatures, which helps find targets hidden by occlusions or camouflage. RGB images provide detailed appearance information. This combination is especially useful in security and military tasks, such as night patrols or search-and-rescue missions. In these situations, tracking heat-emitting objects like people or engines is important, even when visibility is poor.

4.2 Multi-Modal VOT Modalities

Multi-modal target tracking improves performance by integrating various data types, primarily RGB-Depth, RGB-Thermal, RGB-LiDAR, and RGB-Language. Based on the fusion stage, these modalities can be grouped into three strategies: Early, Middle, and Late fusion as shown in Fig. 4.1. Early fusion combines data from different modalities at the input level for joint feature extraction, leveraging

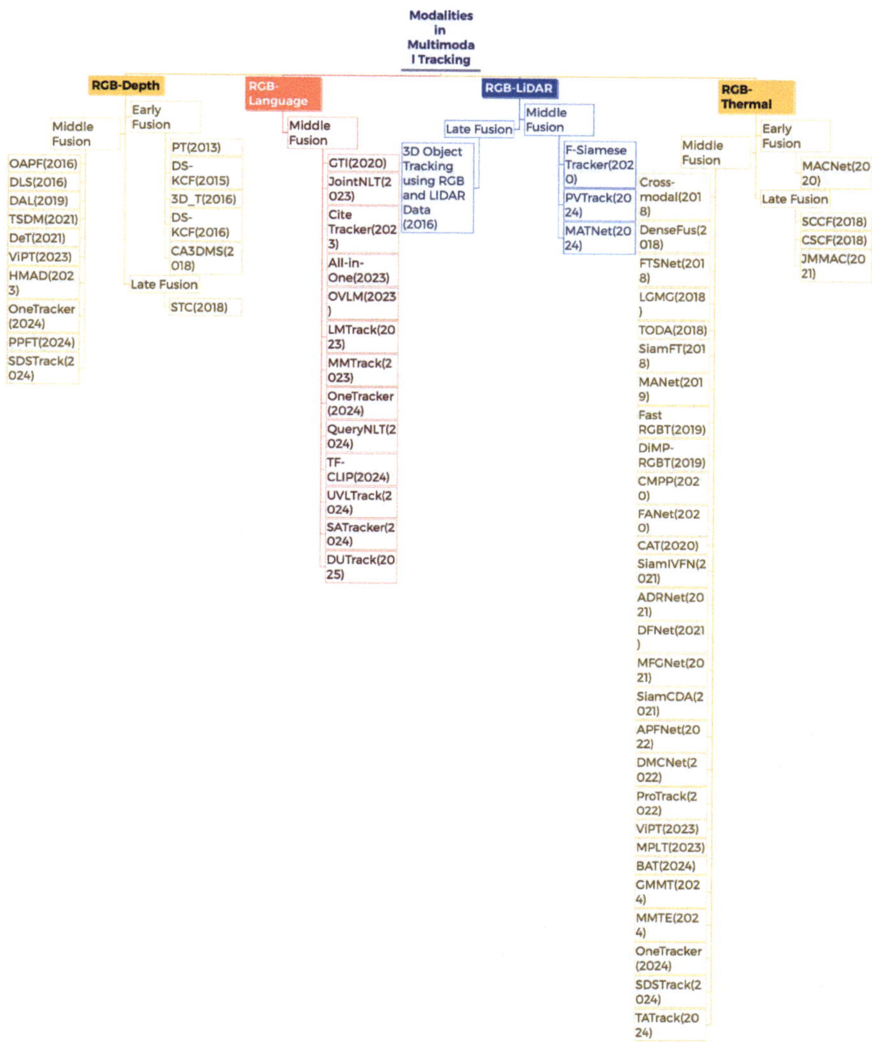

Fig. 4.1 Classification of Multi-modal VOT Methods

all available information from the start. Mid-level fusion extracts features independently from each modality and merges them at the feature level, enriching information while retaining each modality's unique characteristics. Late fusion processes each modality independently and integrates results at the final decision-making stage, ensuring modality independence and maximizing their strengths. Through these approaches, multi-modal target tracking systems achieve a comprehensive understanding of the environment, enhancing accuracy and robustness. Each strategy offers distinct advantages and suits specific application scenarios, providing versatile solutions for efficient tracking. This classification spans the entire process—from data collection to decision-making—and highlights optimization directions for diverse applications, advancing the development and adoption of multi-modal technologies.

4.2.1 RGB-Depth Tracking in VOT

- Early fusion: PT [34] integrates RGB and depth images for feature extraction and constructs a discriminative model using Support Vector Machine (SVM) [6]. DS-KCF [8] proposes an RGB-D visual object tracker based on the fast RGB-only KCF [9] tracker, which incorporates depth information for efficient handling of scale changes, occlusions, and shape changes. CA3DMS [24] explores a 3D extension of the mean-shift tracker. It shows that 3D mean-shift can reliably track objects. This is done by finding distinctive modes in the 3D probability distribution. The method also proposes mechanisms to handle distractions. In addition, it includes strategies to recover from occlusion failures.
- Middle fusion: OAPF [30] introduces an occlusion-aware particle filter framework. It can predict occlusions during tracking. The framework adjusts motion models based on the occlusion state. It also regularizes the effects of features to improve performance. This approach enhances the tracker's resilience to various types of occlusions. It achieves notable improvements in handling both persistent and complex occlusions. DLS [1] proposes a new RGB-D tracking framework. It builds both 2D appearance models and 3D distribution models. This helps the tracker handle large appearance changes. The method breaks tracking into three steps: detection, learning, and segmentation. It also uses an adaptive depth histogram for efficient depth-based localization. As a result, the framework achieves high efficiency, accuracy, and robustness. DAL [14] addresses the limitations of standard RGB-D trackers by performing online 3D target reconstruction to support robust learning of view-specific discriminative correlation filters (DCFs), enhancing performance through accurate spatial support generation and 3D pose estimation for localization after rotation or occlusion. TSDM [63] proposes SiamFT, an RGB-infrared fusion tracking method based on fully convolutional Siamese Networks, which processes visible and infrared images separately before fusing their features adaptively and predicting tracking results in real time. SiamRPN++ [16] proposes an ResNet-driven Siamese tracker with a spatially aware sampling strategy to achieve strict translation invariance and introduces a new model architecture for depth-wise and layer-wise aggregations. DeT [49] introduced a dedicated depth network branch to extract depth features and was the first approach to utilize an independent depth feature extractor. The RGB and depth features were then concatenated and jointly used for tracking.

ViPT [67] as shown in Fig. 4.2, a unified visual prompt-tuning paradigm for downstream multi-modal tracking tasks, does not fully fine-tune an RGB-based tracker combined with an auxiliary-modal fusion branch. Instead, it freezes the entire foundation model and only learns a few modality-specific visual prompts,

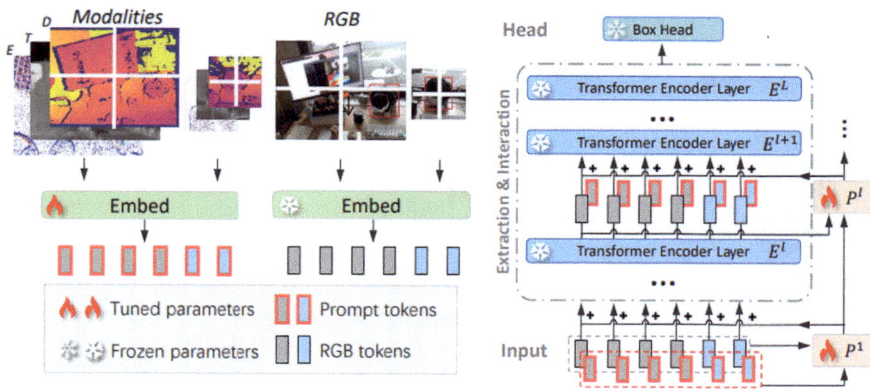

Fig. 4.2 The main architecture diagram of ViPT [67]

which maximally inherits the parameters of the RGB-based model trained at scale. Unlike prompt learning in other single-modal vision tasks, ViPT introduces additional auxiliary-modal inputs into the prompt-tuning process, adapting the foundation model to downstream tasks while simultaneously learning the association between different modalities. Specifically, ViPT inserts several simple and lightweight modality-complementary prompter (MCP) blocks into the frozen foundation model to effectively learn inter-modal complementarities. This approach enables ViPT to leverage the knowledge from large-scale pre-trained models efficiently, achieving state-of-the-art performance across multiple downstream multi-modal tracking tasks while maintaining parameter efficiency.

Given a video with an initial bounding box around the target object, the objective of an RGB tracker is to learn a tracker that predicts the bounding box of the target in subsequent frames. For multi-modal tracking tasks, ViPT extends the model input by introducing an additional spatiotemporally synchronized stream, denoted by (X_{RGB}, X_A), where the subscript A represents other auxiliary modalities such as depth, thermal, or event data. The multi-modal tracker can thus be described as $F_{multi} : (X_{RGB}, X_A, B_0) \rightarrow B$.

Typically, an RGB tracker can be decomposed into two components: $F_{RGB} = \phi \circ f$, where $f : (X_{RGB}, B_0) \rightarrow H_{RGB}$ denotes the feature extraction and interaction function, and $\phi : H_{RGB} \rightarrow B$ represents the bounding box head network that estimates the final result. The backbone of this architecture is a Transformer, where input patches are first embedded and flattened into one-dimensional tokens with positional embeddings, $T = \{t_1, t_2, ..., t_n\}$. These token sequences are then aggregated and passed through standard Transformer encoder layers.

Multi-modal tracking introduces an additional auxiliary modality stream that is temporally synchronized with the RGB stream and spatially aligned. In the proposed method, the two streams, X_{RGB} and X_A, are first passed through a patch embedding layer. Each input stream is mapped and flattened into a one-dimensional latent space. The RGB tokens are denoted as $\mathcal{P}^l_{RGB}$, and the auxiliary tokens are denoted as

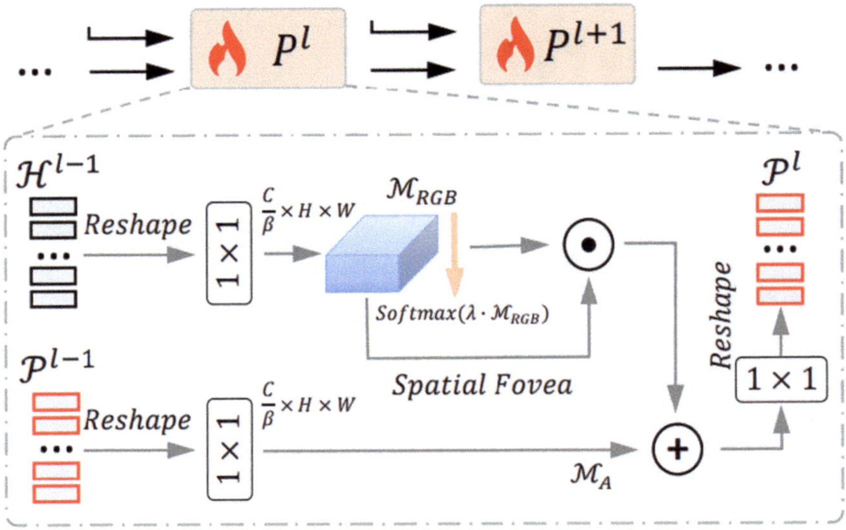

Fig. 4.3 Detailed design of the proposed MCP [67]

$\mathcal{P}^l_A$. These tokens are then fed into the foundation model, where they pass through a cascade modality-complementary prompter (MCP) to generate modality-specific prompts. The learned prompts are added to the original RGB tokens in a residual connection manner:

$$\mathcal{H}^l = \mathcal{H}^l_{RGB} + \mathcal{P}^l, \quad l = 0, 1, \ldots, L-1 \tag{4.1}$$

Some studies explore introducing learnable visual prompts into a frozen pre-trained backbone model by tweaking a small portion of parameters, achieving impressive performance in large-scale vision tasks. More challenging tasks not only require minimizing the gap between upstream and downstream task discrepancies but also effectively utilizing modality information. The MCP module in ViPT is designed to learn prompts for both input streams. The process is described as follows:

$$\mathcal{P}^l = P^l(\mathcal{H}^{l-1}, \mathcal{P}^{l-1}) \tag{4.2}$$

Specifically, through this approach, MCP as shown in Fig. 4.3 disentangles downstream task-specific semantic features while learning inter-modality complementarities and generating more robust prompts. Disentangled representations help balance the utilization of backbone features and learned prompt interactions, aiding in the harmonization of both RGB and auxiliary modalities.

The MCP includes splitting the intermediate feature sequences from the backbone and auxiliary stream into two input branches. The MCP consists of three main steps: (1) Projection to low-dimensional embeddings. (2) Multi-modality internal complementary representation generation and self-adaptation. (3) Projection to original feature dimensions to generate learnable multi-modal prompt representations. Considering that each modality token stream has certain redundant features, the prompt block should contain as few parameters as possible when used in practice. Here, each stream is projected to a low dimension.

In this method, the input dimension is set to 768. All projection heads are set to 12. The projection function g is a simple convolutional layer. Backbone features are first subjected to spatial pooling operations. This operation applies average pooling across all spatial dimensions, followed by applying spatial attention masks to produce enhanced inputs.

$$\mathcal{M}^{e}_{RGB} = \mathcal{M}_{RGB} \odot \mathcal{M}_{fovea} \tag{4.3}$$

$$\mathcal{M}_{fovea} = \left\{ \frac{e^{\mathcal{M}^{[i,j]}_{RGB}}}{\sum e^{\mathcal{M}^{[i,j]}_{RGB}}} \right\} \tag{4.4}$$

Subsequently, additive binding is applied to obtain the combined modality embedding. The learned embedding can be expressed as:

$$\mathcal{P}^{l} = g_3(\mathcal{M}^{e}_{fovea} + \mathcal{M}_A) \tag{4.5}$$

Optimizing the multi-modal tracking model starts with initializing the base model parameters. During the prompt tuning phase, data flow through the entire model but only update the visual prompt learning parameters. Additionally, this method avoids full fine-tuning for both modalities, optimizing the process as follows:

$$\theta_{tuned} = \arg\min_{\theta} \frac{1}{|D|} \sum \mathcal{L}(\phi(\mathcal{H}^{L}), B_{gt}) \tag{4.6}$$

By tweaking only a few parameters for prompt learning, the model converges within a short time frame.

HMAD [48] proposed an attention-based mechanism. It is used for shallow feature extraction and feature distribution fusion. The method integrates both low-level and high-level features across the network. This integration helps to enhance the robustness of tracking. OneTracker [10] employs a Transformer-based approach for RGB-D tracking. The Foundation Tracker, pre-trained on RGB datasets, captures temporal matching. The Prompt Tracker adapts it for RGB-D with Cross Modality Tracking Prompters (CMT) to fuse RGB and depth features and Tracking Task Perception Transformer (TTP) layers for efficient fine-tuning.

PPFT [12] is shown in Fig. 4.4. This chapter proposes a general depth enhancement method based on Polarization Prompt Fusion Tuning (PPFT), leveraging polarization imaging to improve the inaccurate measurements of various depth sensors in complex scenes, such as those involving transparent or reflective objects.

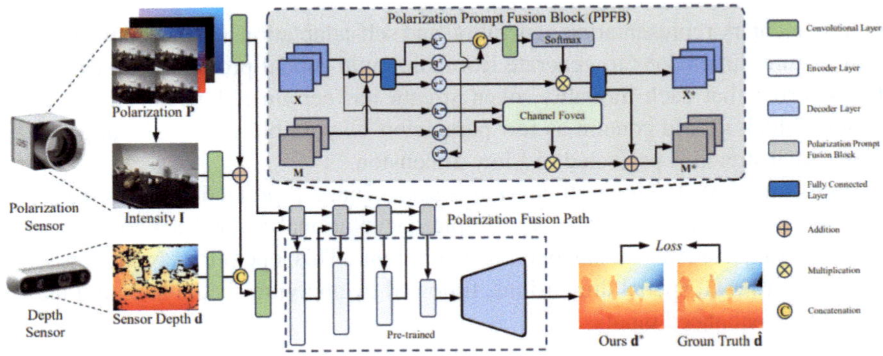

Fig. 4.4 The main architecture diagram of PPFT [12]

The method employs an end-to-end learning framework that integrates polarization data and sensor depth maps to generate high-quality enhanced depth maps, while addressing the challenge of limited polarization dataset size through cross-modal transfer learning. The following details the implementation and technical aspects of the method, including relevant formulas.

The core of the method lies in utilizing polarization data as dense geometric cues to enhance depth sensor performance. Polarization imaging measures the intensity of light passing through a linear polarizer using a Division-of-Focal-Plane (DoFP) camera, capturing geometric information related to surface normals. The light intensity $\mathbf{I}_{\text{pol}}$ is described by the following formula:

$$\mathbf{I}_{\text{pol}} = \mathbf{I}_{\text{un}} \cdot \left(1 + \rho \cdot \cos(2\phi - 2\phi_{\text{pol}})\right), \tag{4.7}$$

where $\mathbf{I}_{\text{un}}$ is the total incident light intensity, ρ is the degree of linear polarization (DoLP), ϕ is the angle of linear polarization (AoLP), and ϕ_{pol} is the polarizer angle. The polarization state is related to the surface normal $\mathbf{n}$ through perspective projection:

$$\begin{cases} \Phi = \mathbf{n} \times \mathbf{v} \times [0, 0, 1]^T, & \text{if diffuse dominant} \\ \Phi = \mathbf{n} \times \mathbf{v} \times \mathbf{v} \times [0, 0, 1]^T, & \text{if specular dominant} \end{cases} \tag{4.8}$$

$$\phi = \arctan\left(\frac{\Phi_y}{\Phi_x}\right), \tag{4.9}$$

where $\Phi = [\Phi_x, \Phi_y, \Phi_z]^T$, and $\mathbf{v} \in \mathbb{R}^3$ is the viewing direction. Due to the challenge of directly deriving surface normals, the method adopts a learning strategy, constructing polarization guidance data $\mathbf{P} = [\mathbf{I}; \phi; \rho; \mathbf{V}]$, comprising the intensity image $\mathbf{I}$, polarization angle ϕ, polarization degree ρ, and viewing direction $\mathbf{V}$. The DoFP camera captures images at four polarization angles (0, $\pi/4$, $\pi/2$, $3\pi/4$), recovering ϕ and ρ based on the formula.

Depth enhancement is achieved through a neural network f, which, given the raw sensor depth map $\mathbf{d}$ and polarization data $\mathbf{P}$, predicts the enhanced depth $\hat{\mathbf{d}}$:

$$\hat{\mathbf{d}} = f(\mathbf{P}, \mathbf{d}; \theta), \tag{4.10}$$

where θ denotes the network parameters. Training employs a combination of L_1 and L_2 losses:

$$\mathcal{L}(\mathbf{d}^*, \hat{\mathbf{d}}) = \frac{1}{|\mathbf{M}_{\text{depth}}|} \sum_{m \in \mathbf{M}_{\text{depth}}} \left(\|\mathbf{d}_m^* - \hat{\mathbf{d}}_m\|_1 + \|\mathbf{d}_m^* - \hat{\mathbf{d}}_m\|_2^2 \right), \tag{4.11}$$

where $\mathbf{d}^*$ is the ground truth depth, and $\mathbf{M}_{\text{depth}}$ is the mask for valid depth pixels.

To address the challenge of limited polarization dataset size, the method introduces the PPFT strategy, leveraging a CompletionFormer model pretrained on large-scale RGB-D datasets (e.g., NYU-Depth V2) to enhance generalization via cross-modal transfer learning. The core of PPFT is the Polarization Prompt Fusion Block (PPFB), which progressively fuses polarization features across multiple encoder layers through parallel branches, mitigating modality misalignment between RGB and polarization. PPFB consists of two branches: the input feature update branch uses a spatial attention mechanism to update input features $\mathbf{X}$ with polarization prompts $\mathbf{M}$, producing fused features $\mathbf{X}^*$:

$$[\mathbf{k}^x; \mathbf{q}^x; \mathbf{v}^x] = [\mathbf{M}; \mathbf{X}] \times [\mathbf{W}_k^x; \mathbf{W}_q^x; \mathbf{W}_v^x], \tag{4.12}$$

$$[\mathbf{X}^*; \mathbf{M}^x] = \text{FC}_d \left(\lambda \cdot \text{softmax}(\text{FC}([\mathbf{q}^x; \mathbf{k}^x])) \odot \mathbf{v}^x \right), \tag{4.13}$$

where $\mathbf{W}_k^x, \mathbf{W}_q^x, \mathbf{W}_v^x \in \mathbb{R}^{2C \times C}$ are learnable parameters, FC_d is a fully connected layer with dropout, and $\odot$ denotes element-wise multiplication. The polarization prompt learning branch updates the prompt $\mathbf{M}^*$ via a channel fovea operation:

$$\mathbf{k}^m, \mathbf{q}^m, \mathbf{v}^m = \mathbf{X}, \mathbf{M}, \mathbf{k}^x, \tag{4.14}$$

$$[s_q; s_k] = \frac{1}{H \times W} \sum_{i=1}^{H} \sum_{j=1}^{W} (\text{FC}(\mathbf{M} + \mathbf{X})), \tag{4.15}$$

$$\mathbf{M}^* = \mathbf{M}^x + \text{FC}\left((s_q \mathbf{Q}^m + s_k \mathbf{K}^m) \odot \mathbf{V}^m \right). \tag{4.16}$$

The channel fovea operation extracts channel statistics s_q, s_k via global average pooling, enhancing the channel-wise correlation of polarization prompts.

The network first processes polarization data, depth maps, and other modalities through three convolutional encoders, generating modality-specific embeddings that are fused into a multi-modal embedding and fed into PPFB. PPFB updates polarization prompts and features at each encoder layer, ensuring sustained influence of polarization information in deeper layers, with the final enhanced depth map produced by a decoder. PPFT effectively utilizes pretrained weights through deep polarization data fusion, achieving robust enhancement across various depth sensors.

- Late fusion: STC [44] performs fusion after tracking by integrating weighted estimates of the target's final position from both RGB and depth modalities, thus carrying out the fusion step posterior to feature extraction.

4.2.2 RGB-Language Tracking in VOT

- Middle Fusion: GTI [51] followed a "grounding-then-tracking" framework and combined both stages using scores related to region accuracy and template quality. All-in-One [54] proposed a unified transformer backbone for vision–language tracking, integrating feature extraction and interaction by mixing raw vision and language signals into a joint representation, and further enhancing learning efficiency with a multi-modal alignment module. OVLM [55] introduced a one-stream vision–language memory network for object tracking. It combines visual and language information to create a stable target model. The method uses linguistic information to remove irrelevant tokens. It also selects high-quality samples to improve tracking performance. CiteTracker [23] incorporated a text generation component that produces a textual description of the tracked object, which is then used for state estimation. MMTrack [64] proposed a simple and flexible vision–language tracking pipeline that reformulates VL tracking as a token generation task, serializing language and bounding boxes into tokens to directly predict target coordinates autoregressively. LMTrack [27] introduced a language-guided attention mixer for visual tracking, which adaptively guides visual attention using natural language and corrects visual bias through linguistic information. JointNLT [66] proposed a unified framework integrating multi-source relation modeling for visual grounding and tracking, using natural language specifications, and incorporated a semantics-guided temporal modeling module to enhance adaptability to target appearance variations. SATracker [7] presents a novel Vision–Language Tracking (VLT) tracker with a Synchronous Learning Backbone (SLB) that includes a Target Enhance Module (TEM) and a Semantic-Aware Module (SAM) to ensure synchronous multi-modal feature extraction and interaction and introduces a dense matching loss to strengthen multi-modal representation learning, achieving superior performance on VLT datasets. UVLTrack [29] all proposed to utilize contrastive learning strategy to minimize the distances between positive vision–language token pairs.

As shown in Fig. 4.5, QueryNLT [33] processes a search image $I_s \in \mathbb{R}^{3\times H_s \times W_s}$ and a language description $\mathcal{D}$ to derive discriminative features. A Swin-Transformer, limited to its first three stages for efficiency, serves as the visual backbone, outputting a search feature $f_s \in \mathbb{R}^{N_s \times C}$ (with N_s features and $C = 256$) after channel adjustment. For text, a RoBERTa model with a projection layer extracts features $f_l \in \mathbb{R}^{L\times C}$, where L is the text length and $C = 256$. A bi-attention

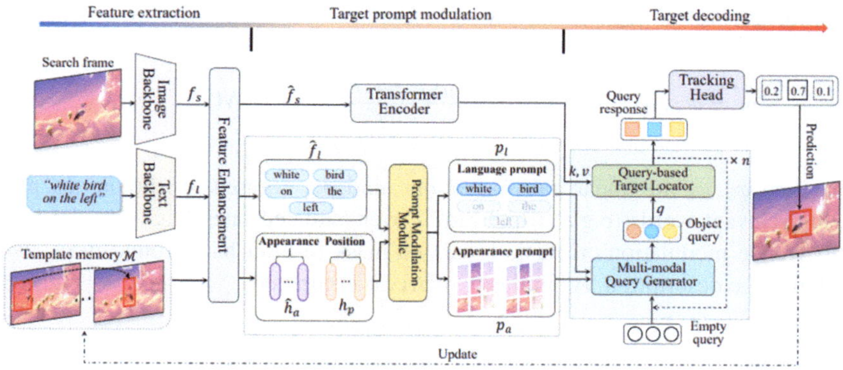

Fig. 4.5 The main architecture diagram of QueryNLT [33]

mechanism enhances these features by cross-referencing the search image with the text and historical appearance features h_a from a template memory $\mathcal{M} = \{h_a, h_p\}$, which stores appearance and positional data from prior frames. This enhancement is expressed as:

$$\hat{f}_s = \omega_s \left(f_s + \text{softmax}\left(\frac{f_s f_l^T}{\sqrt{C}}\right) f_l + \text{softmax}\left(\frac{f_s h_a^T}{\sqrt{C}}\right) h_a \right), \tag{4.17}$$

$$\hat{f}_l = \omega_l \left(f_l + \text{softmax}\left(\frac{f_l f_s^T}{\sqrt{C}}\right) f_s \right), \tag{4.18}$$

$$\hat{h}_a = \omega_a \left(h_a + \text{softmax}\left(\frac{h_a f_s^T}{\sqrt{C}}\right) f_s \right), \tag{4.19}$$

where $\omega_s, \omega_l, \omega_a$ are linear layers, and softmax computes attention weights to align features across modalities, preserving positional features h_p to maintain motion information.

The prompt modulation module ensures accurate target cues by addressing misalignments in language descriptions and background clutter in visual templates. Language modulation uses motion cues from h_p and appearance cues from $\hat{h}_a$ to re-weight the text feature $\hat{f}_l$. Self-attention is applied to $\hat{h}_a$ and h_p, followed by the addition of identifier vectors $v_a, v_p \in \mathbb{R}^C$ to differentiate appearance and motion. The template memory is processed as:

$$\mathcal{M} = \left[\theta_a(\hat{h}_a) + [v_a]^{N_a}, \theta_p(h_p) + [v_p]^{N_p}\right], \tag{4.20}$$

where θ_a, θ_p are self-attention operations, $[\cdot]^n$ duplicates vectors n times, and N_a, N_p denote the number of appearance and position features. A multi-head cross-attention generates the language prompt:

$$p_l = \varphi_{\text{agg}}\left(\hat{f}_l + \text{softmax}\left(\frac{\varphi_q(\hat{f}_l)\varphi_k(\mathcal{M})^T}{\sqrt{C}}\right)\varphi_v(\mathcal{M})\right), \tag{4.21}$$

where $\varphi_q, \varphi_k, \varphi_v$ are linear transformations, and φ_{agg} aggregates the output, prioritizing relevant words (e.g., "white bird") over outdated ones (e.g., "on the left"). Appearance modulation creates a binarized mask to filter background in $\hat{h}_a$ using the language description. A similarity matrix is computed:

$$A_{l\mapsto a} = \text{softmax}\left(\frac{\delta_a(\hat{h}_a)\delta_l(\hat{f}_l)^T}{\sqrt{C}}\right), \tag{4.22}$$

where δ_a, δ_l are linear layers, and $A_{l\mapsto a} \in \mathbb{R}^{N_a\times L}$ maps pixels to words. To focus on the target, subjecthood scores are calculated using Gumbel-Softmax:

$$A_{obj}^i = \frac{\exp\left(\mathbf{W}_{obj}\hat{f}_l^i + \gamma_i\right)}{\sum_{j=1}^{L}\exp\left(\mathbf{W}_{obj}\hat{f}_l^j + \gamma_j\right)}, \tag{4.23}$$

where $\mathbf{W}_{obj} \in \mathbb{R}^{1\times C}$ is a projection weight, and γ_i, γ_j are Gumbel samples. A one-hot operation yields:

$$\hat{A}_{obj} = \text{one-hot}(A_{obj}^{\text{argmax}}) + A_{obj} - \text{sg}(A_{obj}), \tag{4.24}$$

where sg is the stop-gradient operator. The appearance prompt is:

$$M_{obj} = A_{l\mapsto a} \times \hat{A}_{obj}, \quad p_a = \hat{h}_a \times M_{obj}, \tag{4.25}$$

where $M_{obj} \in \mathbb{R}^{N_a\times 1}$ masks non-target pixels, enhancing the visual prompt.

The target decoding module unifies language and appearance matching using a transformer-based architecture inspired by Deformable-DETR. A query generator creates an object-aware query vector q_{obj} from an initial empty vector q_{init}:

$$q_{obj} = \psi\left(q_{\text{init}} + \alpha\sum_{i=1}^{N_a} a^i p_a^i + (1-\alpha)\sum_{j=1}^{N_l} b^j p_l^j\right), \tag{4.26}$$

where ψ is a linear layer, α balances modalities (set to 0 for language-only initialization), and attention weights a^i are:

$$a^i = \frac{\exp\left(\mathbf{W}_a p_a^i \mathbf{W}_q q_{\text{init}}\right)}{\sum_{k=1}^{N_a}\exp\left(\mathbf{W}_a p_a^k \mathbf{W}_q q_{\text{init}}\right)}, \tag{4.27}$$

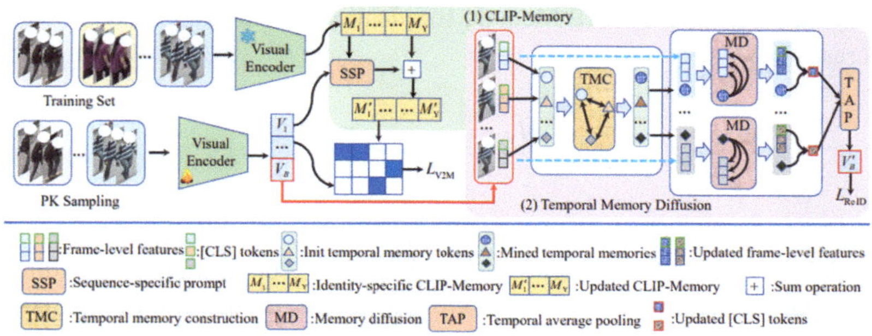

Fig. 4.6 The main architecture diagram of TF-CLIP [52]

with $\mathbf{W}_a, \mathbf{W}_q \in \mathbb{R}^{1\times C}$. The query q_{obj} performs cross-attention with the enhanced search feature $\hat{f}_s$ (as Key and Value) in a transformer encoder, refining the query over stacked decoder layers to produce a response r containing the target state. Multiple query vectors (optimally five, per ablation studies) capture diverse prompt interpretations, with classification and regression heads predicting scores and bounding boxes, supervised by cross-entropy, L1, and GIoU losses. The highest-scoring box is selected, and the template memory is updated with ROI-aligned target features and positional data, ensuring robust tracking across dynamic scenes.

As shown in Fig. 4.6, TF-CLIP [52] integrates a Sequence-Specific Prompt (SSP) module to dynamically update CLIP-Memory, ensuring adaptability to appearance variations, and employs contrastive, triplet, and cross-entropy losses for optimization, making it robust for fine-grained person re-identification.

The CLIP-Memory module replaces the text branch by computing identity-specific sequence features from video sequences. Given a training set $\mathcal{D} = \{(x_i, y_i)\}_{i=1}^{N}$ with N tracklets and Y identities, for an identity y_i, the pre-trained CLIP visual encoder processes video sequences $\{I_t\}_{t=1}^{T}$, where each frame $I_t \in \mathbb{R}^{H\times W\times 3}$ is divided into $N_p = \frac{H}{P}\times\frac{W}{P}$ patches $\{I_{t,i}\}_{i=1}^{N_p} \in \mathbb{R}^{P^2\times 3}$. These patches are projected into token embeddings via a linear layer $E_{\text{emb}} \in \mathbb{R}^{3P^2\times D}$, and a learnable [CLS] token $CLS_t \in \mathbb{R}^D$ is added, forming the input for frame t:

$$z_t^{\text{Patch}} = [CLS_t, E_{\text{emb}}^T I_{t,1}, \ldots, E_{\text{emb}}^T I_{t,N_p}] + e^{\text{sp}}, \tag{4.28}$$

where e^{sp} is the spatial position embedding, and D is the token dimension. The visual encoder $\mathcal{V}^*$ (with frozen parameters) processes these embeddings to yield frame-level features:

$$z_t = \mathcal{V}^*(z_t^{\text{Patch}}), \tag{4.29}$$

where $z_t \in \mathbb{R}^{(N_p+1)\times D}$. The [CLS] token $z_{t,0}$ is projected via a Visual Project (VP) layer, $f_t = VP(z_{t,0})$, with $VP = W_{VP} \in \mathbb{R}^{D\times d}$, to obtain the frame representation f_t. Temporal Average Pooling (TAP) aggregates frame features into a sequence-level feature v_a, and the CLIP-Memory for identity y_i is the average of all sequence features:

$$M_{y_i} = \frac{1}{N_i} \sum_{v_a \in y_i} v_a, \tag{4.30}$$

where N_i is the number of sequences for y_i, forming $M \in \mathbb{R}^{Y \times d}$. To address the limitation of a fixed CLIP-Memory, which may overfit to training identities, the SSP module updates it online using a cross-attention mechanism. For a mini-batch $\{I^b\}_{b=1}^{B=P\times K}$ with P identities and K sequences per identity, sequence-level features $V = [v_{y=1}^1, \ldots, v_{y=P}^K] \in \mathbb{R}^{PK\times d}$ are computed. SSP uses M as the query and V as key and value in a Transformer decoder with $N = 2$ blocks, each containing Multi-Head Cross-Attention (MHCA) and a Feed-Forward Network (FFN):

$$\mathcal{P}_{y=1,\ldots,y=P} = \text{Transdecoder}(\text{query} = M, \text{key} = V, \text{value} = V), \tag{4.31}$$

updating the CLIP-Memory via a residual connection:

$$M'_{y=1,\ldots,y=P} = \mathcal{P}_{y=1,\ldots,y=P} + M_{y=1,\ldots,y=P}, \tag{4.32}$$

yielding $M' \in \mathbb{R}^{Y\times d}$. This dynamic update enhances adaptability to sequence-specific variations, reducing overfitting.

The TMD module captures temporal information through Temporal Memory Construction (TMC) and Memory Diffusion (MD). For a sequence $I^B = \{I_1^B, \ldots, I_T^B\}$, TMC takes frame-level features $z_t \in \mathbb{R}^{(N_p+1)\times D}$ and constructs a memory token for each frame:

$$s_t = \theta\left(\frac{1}{N_p + 1}\sum_{i=0}^{N_p} z_{t,i}\right), \tag{4.33}$$

where θ is a linear projection. A Multi-Head Self-Attention (MHSA) over the set of memory tokens $S = \{s_1, \ldots, s_T\}$ extracts temporal relations:

$$S' = \text{MHSA}(\text{LN}(S)) + S, \tag{4.34}$$

where $S' \in \mathbb{R}^{T\times D}$ are temporal-aware memory tokens, and LN denotes Layer Normalization. MD diffuses these tokens to frame features by concatenating $s'_t \in \mathbb{R}^{1\times D}$ with z_t, forming $[z_t, s'_t] \in \mathbb{R}^{(N_p+2)\times D}$, and applying MHSA and FFN:

$$[\hat{z}_t, \hat{s}'_t] = \text{FFN}(\text{MHSA}([z_t, s'_t])) + [z_t, s'_t], \tag{4.35}$$

where $\hat{z}_t \in \mathbb{R}^{(N_p+1)\times D}$. Frame-level features are aggregated:

$$\bar{z}_t = \frac{1}{N_p+2}\left(\sum_{i=0}^{N_p} \hat{z}_{t,i} + \hat{s}'_t\right), \tag{4.36}$$

and TAP yields the sequence-level feature:

$$\hat{v} = \mathrm{TAP}([\bar{z}_1, \ldots, \bar{z}_T]). \tag{4.37}$$

The framework is optimized using a video-to-memory contrastive loss L_{V2M}, triplet loss L_{tri}, and label-smoothed cross-entropy loss L_{ce}. For a batch with sequence features $V = [v_1, \ldots, v_B]$ and updated CLIP-Memory M', the contrastive loss is:

$$L_{V2M}^{(y_i)} = -\frac{1}{|P(y_i)|}\sum_{p\in P(y_i)} \log \frac{\exp(\langle v_p, M'_{y_i}\rangle)}{\sum_{j=1}^{B}\exp(\langle v_j, M'_{y_i}\rangle)}, \tag{4.38}$$

where $\langle\cdot\rangle$ is cosine similarity, and $P(y_i)$ denotes positive samples for identity y_i. The total loss is:

$$L_{\text{total}} = L_{V2M} + L_{\text{tri}} + L_{\text{ce}}, \tag{4.39}$$

ensuring robust feature learning. During testing, the original feature $v \in \mathbb{R}^{1\times 512}$ and aggregated feature $\hat{v} \in \mathbb{R}^{1\times 768}$ are concatenated, using Euclidean distance for ranking, achieving discriminative embeddings for video-based ReID.

DUTrack [22] presented a vision–language tracking framework that dynamically updates multi-modal references using a Dynamic Template Capture Module and a Dynamic Language Update Module to maintain consistency with the target's state, achieving state-of-the-art performance on multiple benchmarks.

4.2.3 RGB-LiDAR Tracking in VOT

- Middle Fusion: The F-Siamese Tracker [69] utilizes a Siamese network to generate 2D region of interest (ROI) proposals from RGB images. It then integrates these ROI proposals with point cloud data. This combined information is subsequently fed into a second Siamese network, which outputs 3D bounding boxes for object tracking. PVTrack [47], as shown in Fig. 4.7, operates through a three-part pipeline: multi-modal feature extraction and fusion, Siamese Point-Video Transformer, and 3D Human-ware Proposal and Verification. The feature extraction module employs lightweight Transformer-based backbones to process point clouds and RGB video frames separately, addressing the computational

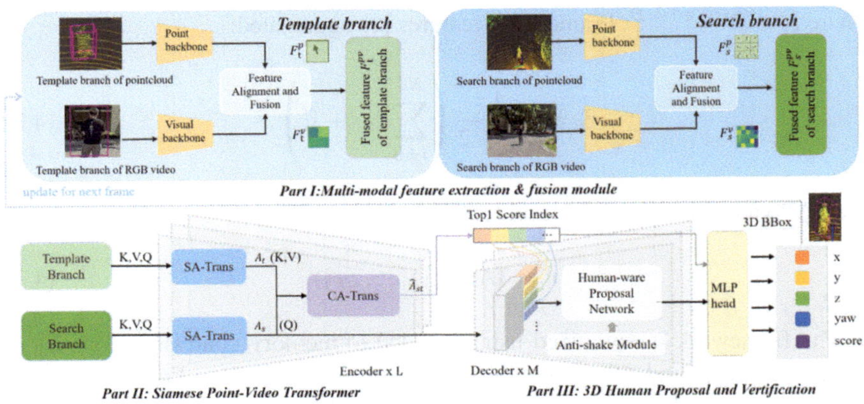

Fig. 4.7 The main architecture diagram of PVTrack [47]

inefficiency of traditional local descriptor networks like PointNet++. For point clouds, a modified Point-MAE extracts features, producing a point feature set $F^p = \{f_i^p\}_{i=1}^K \in \mathbb{R}^{K \times C_1}$, where K is the number of point feature groups after down-sampling, and C_1 denotes the LiDAR channel dimension. For video frames, MobileFormer extracts features, yielding a feature map $F^v = \{f_{(i,j)}^v\} \in \mathbb{R}^{H' \times W' \times C_2}$, where $H' \times W'$ is the spatial dimension of the feature map, and C_2 represents the RGB channel dimension. To align and fuse these heterogeneous modalities, the Point-Video Feature Alignment and Fusion (PVAF) operation is introduced. PVAF first applies farthest point sampling (FPS) to select K point groups with center points $\{c_i\}_{i=1}^K$. The point features $\{f_i^p\}_{i=1}^K$ are then interpolated back to the original N points $\{p_i'\}_{i=1}^N$ using inverse distance weights, as defined by:

$$p_i' = \sum_{j=1}^{K} w_j f_j^p, \quad w_j = \frac{\frac{1}{\|p_i - c_j\|_2 + \epsilon}}{\sum_{k=1}^{K} \frac{1}{\|p_i - c_k\|_2 + \epsilon}}, \tag{4.40}$$

where ϵ is a small constant to avoid division by zero. The interpolated point features $p' \in \mathbb{R}^{N \times C_1}$ are projected into 2D coordinates $\tilde{p} \in \mathbb{R}^{H' \times W' \times C_3}$ using LiDAR-camera calibration, aligning them with the video feature map f^v. A multilayer perceptron (MLP) $\chi(\cdot)$ extracts interaction information, and the features are concatenated to form the fused point-video feature $F^{pv} = \{f_{(i,j)}^{pv}\} \in \mathbb{R}^{M \times C}$, where $M = H' \times W'$, $C = C_2 + C_3$, and:

$$f_{(i,j)}^{pv} = \chi^p \left(f_{(i,j)}^p \right) \oplus \chi^v \left(f_{(i,j)}^v \right). \tag{4.41}$$

This fusion enables PVTrack to capture both geometric and appearance cues, enhancing tracking accuracy in scenarios like distinguishing targets during close encounters based on clothing color.

The Siamese Point-Video Transformer processes the fused features through a hierarchical attention mechanism, comprising Self-Attention Transformer (SA-Trans) and Cross-Attention Transformer (CA-Trans) modules. SA-Trans fuses adjacent point and video features within each branch (template and search) by computing attention maps. For input features f_s^{pv} (search branch) and f_t^{pv} (template branch), the self-attention is formulated as:

$$\mathrm{SA}(\mathbf{Q}, \mathbf{K}, \mathbf{V}) = \phi\left(\mathbf{Q} \cdot \mathrm{softmax}(\mathbf{Z}) \cdot (W_v \mathbf{V})\right), \tag{4.42}$$

where $\mathbf{Z} = \tilde{\mathbf{Q}} \cdot \tilde{\mathbf{K}}^T = \frac{W_q \mathbf{Q}}{\|W_q \mathbf{Q}\|_2} \cdot \frac{W_k \mathbf{K}}{\|W_k \mathbf{K}\|_2}$, ϕ denotes a linear layer with ReLU, and W_q, W_k, W_v are projection matrices. This yields attention maps $A_s = \mathrm{SA}(f_s^{pv}, f_s^{pv}, f_s^{pv})$ and $A_t = \mathrm{SA}(f_t^{pv}, f_t^{pv}, f_t^{pv})$. CA-Trans then embeds template features into the search region to compute cross-correlation, producing a refined attention map:

$$\hat{A}_{st} = \mathrm{CA}(A_s, A_t, A_t), \tag{4.43}$$

where CA-Trans follows a similar architecture but uses template features as keys and values to enhance search features, enabling coarse-to-fine similarity learning.

The 3D Human-ware Proposal Network (HPN) and Anti-shake Proposal Compensation module finalize the tracking process. HPN takes the refined attention map $\hat{A}_{st}^L$ from the last encoder layer, concatenates template and corresponding search features, and generates coarse proposals using a P2B-like network. Human-specific priors (width: 10–40 cm, height: 1.0–2.0 m, length: 20–60 cm) filter proposals to accelerate processing and improve accuracy. The Anti-shake module addresses the violent shaking of quadruped robots by expanding the candidate region based on the target's velocity, calculated as $S = \max(\{S_{t-1}, \|X_t - X_{t-1}\|_2\})$. When the maximum response r^t falls below a threshold $\Omega \times \mathrm{mean}(r^t)$, the module adjusts the bounding box dimensions iteratively:

$$W^n = n \cdot S + 2 \cdot bb(1), \quad H^n = n \cdot S + 2 \cdot bb(2), \quad L^n = n \cdot S + 2 \cdot bb(3), \tag{4.44}$$

where bb represents the previous frame's bounding box dimensions, and n is the iteration counter. This process extracts points within the expanded region to resume tracking. The final prediction is selected via an MLP head, optimizing a loss function combining binary cross-entropy for classification and Smooth-L1 for regression:

$$\mathcal{L}(\hat{y}^t \mid \omega^c, y^t) = \lambda_{cls} \mathcal{L}_{cls} + \lambda_{reg} \sum \mathcal{L}_{reg}(\mathbf{b}, \hat{\mathbf{b}}), \tag{4.45}$$

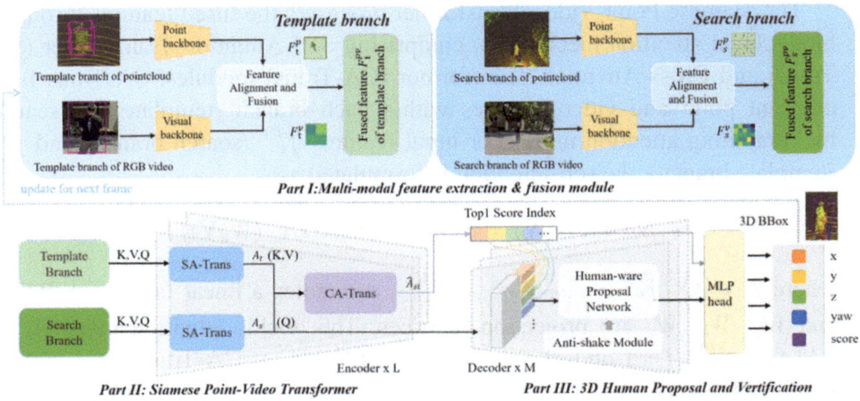

Fig. 4.8 The main architecture diagram of MATNet [46]

where $\mathbf{b} = \{x, y, z, w, h, l, \theta\}$ represents the 3D bounding box parameters, and $\lambda_{cls} = \lambda_{reg} = 0.5$. The total loss over T frames is:

$$\mathcal{L}^{\text{total}}(\hat{Y} \mid \omega, Y) = \frac{\sum_{t=1}^{T} \mathcal{L}(\hat{y}^t \mid \omega, y^t)}{T}. \tag{4.46}$$

MATNet [46], as shown in Fig. 4.8, comprises three main components: a Multi-modal Feature Extraction Module, a Coarse-level Transformer for 2D-3D matching, and a Fine-level Matching Module for precise 3D bounding box generation. The feature extraction module processes RGB video frames $V \in \mathbb{R}^{H \times W \times 3}$ and LiDAR point clouds $P \in \mathbb{R}^{N \times 3}$ using a single-stream network to reduce computational complexity, unlike traditional Siamese trackers that employ separate template and search branches. For video inputs, a Feature Pyramid Network (FPN) extracts multi-level features, producing coarse features $F_c^V = [z_c^V, x_c^V] \in \mathbb{R}^{(N_c+N_s)\times D}$ at 1/16 resolution and fine features $F_f^V = [z_f^V, x_f^V] \in \mathbb{R}^{(N_c+N_s)\times D}$ at 1/4 resolution, where z_c^V, x_c^V are template and search frame features, and D is the feature dimension. For point clouds, a KPConv-FPN performs multilayer down-sampling to generate five point groups $P^k \in \mathbb{R}^{N_k \times 3}$ (k=1,2,3,4,5) with varying resolutions, processed by five encoder layers and one decoder layer to yield coarse features $F_c^P = [z_c^P, x_c^P]$ and fine features $F_f^P = [z_f^P, x_f^P]$. To enhance efficiency for robotic deployment, MATNet employs adapter-based tuning, integrating lightweight adapters into the FPN block. Each adapter consists of two fully connected layers with a GELU activation and a residual connection, formulated as:

$$\overline{\mathbf{H}}^{(l)} = \mathbf{H}^{(l-1)} + \text{Adapter}(\text{CNN}(\mathbf{H}^{(l-1)})), \tag{4.47}$$

$$\mathbf{H}^{(l)} = \overline{\mathbf{H}}^{(l)} + \text{MLP}(\text{LN}(\overline{\mathbf{H}}^{(l)})) + r \cdot \text{Adapter}(\text{LN}(\overline{\mathbf{H}}^{(l)})), \tag{4.48}$$

where $\mathbf{H}^{(l-1)}$, $\mathbf{H}^{(l)}$ are the input and output features of the l-th FPN block, LN denotes layer normalization, and r is a scaling factor. This approach freezes pre-trained weights, reducing the trainable parameters by 2–3 times and training time from 15 hours (100+ epochs) to 4 hours (45 epochs), enabling efficient fine-tuning on robotic datasets.

The Coarse-level Transformer module performs 2D–3D matching using a seq2seq Transformer framework with self-attention and cross-attention mechanisms. It processes coarse features $[F_c^V, F_c^P]$ with position embeddings $[P^1, P^5]$ to capture contextual dependencies. The attention mechanism computes similarity between query $\mathbf{Q}$, key $\mathbf{K}$, and value $\mathbf{V}$, defined as:

$$A_c = \text{Attention}(\mathbf{Q}, \mathbf{K}, \mathbf{V}) = \phi\left(\mathbf{Q} \cdot \text{softmax}(\mathbf{Z}) \cdot (W_v \mathbf{V})\right), \tag{4.49}$$

$$\mathbf{Z} = \tilde{\mathbf{Q}} \cdot \tilde{\mathbf{K}}^T = \frac{W_q \mathbf{Q}}{\|W_q \mathbf{Q}\|_2} \cdot \frac{W_k \mathbf{K}}{\|W_k \mathbf{K}\|_2}, \tag{4.50}$$

where ϕ is a linear layer with ReLU, and W_q, W_k, W_v are projection matrices. To align 2D and 3D features, a transition matrix $S \in \mathbb{R}^{(1/k)^2 HW \times N} = \langle \text{normalize}(A_c^V), \text{normalize}(A_c^P) \rangle$ is computed, extended to $\hat{S}$ with a row for unmatched points. A coarse-level confidence matrix M_c is derived using:

$$M_c = \left\{ (i_c^V, i_c^P, i_c^V) \mid \hat{S} = \max(\text{Softmax}(\hat{S}(\cdot, i/t))), \hat{S} > th_c \right\}, \tag{4.51}$$

where t is a temperature parameter, and th_c is a threshold, selecting candidate regions for further refinement.

The Fine-level Matching Module refines these candidates using part-aware and size-aware processing. For 2D video, a window of size $w \times w$ is cropped around the highest-confidence regions in M_c, concatenated with F_f^V to update track queries. For 3D point clouds, k-nearest neighbors (k-NN) sampling extracts a local point cloud P^{loc} from P^3 based on Euclidean distance. A Human-ware Correlation Head, adapted from prior work, filters candidates by discarding over-sized boxes, ensuring geometric discriminability. The Anti-shake Refinement strategy addresses platform shaking by computing target velocity over the last 10 frames:

$$S = \max\left\{ 5, \sum_{i=1}^{t} \|X_i - X_{i-1}\|_2 \right\}, \tag{4.52}$$

where X_i is the 3D location of the target. If the response score r^t falls below threshold Ω, the candidate region is expanded:

$$W^n = n \cdot S + 2 \cdot bb(1), \quad H^n = n \cdot S + 2 \cdot bb(2), \quad L^n = n \cdot S + 2 \cdot bb(3), \tag{4.53}$$

where bb denotes the previous frame's bounding box dimensions, and n is the iteration counter. Points within the expanded region $I_t^{W^n \times L^n \times H^n}$ are extracted, and the maximum response r^n is computed. If r^n exceeds Ω, tracking resumes; otherwise, the process iterates to the next frame. A three-layer MLP selects the highest-scoring proposal as the final 3D bounding box, optimized using a loss function:

$$L = \frac{1}{T}\left(\lambda^{cls}\sum_{i=1}^{T} L_i^{cls} + \lambda^{reg}\sum L_i^{reg}(bb)\right), \tag{4.54}$$

where L^{cls} is cross-entropy loss, L^{reg} is Huber loss, $\lambda^{cls} = 0.4$, and $\lambda^{reg} = 0.6$. MATNet's robustness is demonstrated across challenging scenarios, including occlusion, rainy days, and dark tunnels, with a 20% accuracy improvement from the Anti-shake Refinement and superior performance on the PVT-3D dataset, making it a significant advancement for robotic 3D human tracking.

- Late Fusion: Alireza et al. [2] employed two separate mean-shift localizers to estimate the target's position in both 2D image space and 3D physical space. The information from these two modalities was then integrated using a Kalman filter for improved localization accuracy.

4.2.4 RGB-Thermal Tracking in VOT

- Early Fusion: MACNet [56] introduces a novel tracking framework that leverages attention mechanisms and adversarial learning to achieve robust visual tracking performance across various challenging scenarios, such as occlusion, deformation, and low illumination.
- Middle Fusion: Cross-modal [18] proposes a cross-modal ranking algorithm to suppress background effects for robust RGB-T tracking by integrating soft cross-modality consistency and optimal query learning. LGMG [20] proposes a novel local–global multi-graph descriptor learning approach for RGB-T object tracking to suppress background effects and prevent model drifting. DenseFuse [21] introduces a deep learning architecture for infrared and visible image fusion, featuring convolutional layers, a fusion layer, and dense blocks. FTSNet [19] proposes a new RGB-T tracking method. It uses a two-stream ConvNet and a FusionNet to extract features from RGB and thermal modalities. These networks fuse the features adaptively. The method achieves state-of-the-art performance. It selects discriminative features and predicts object locations. A multi-channel correlation filter is used for accurate localization. TODA [43] proposes a novel multi-spectral tracking method using collaborative sparse correlation filters to enhance tracking accuracy in adverse illumination conditions while maintaining real-time performance. MANet [25] introduces a Multi-Adapter convolutional Network for RGBT tracking that jointly performs modality-shared, modality-specific, and

instance-aware feature learning using three types of adapters, achieving outstanding performance on RGBT tracking benchmarks. Fast RGBT [53] proposes a novel RGB-T tracking algorithm using low-rank-constrained collaborative correlation filters to leverage interdependency between modalities, achieving state-of-the-art accuracy and efficiency. DiMP-RGBT [57] introduces an end-to-end RGB-T tracking framework based on DiMP, exploring various fusion mechanisms (pixel-level, feature-level, and response-level). CMPP [36] proposes a Cross-Modal Pattern-Propagation tracking framework that diffuses instance patterns across RGB-T data in both spatial and temporal domains by leveraging cross-modal correlations and temporal continuity. FANet [68] achieves robust RGBT tracking by aggregating hierarchical features within each modality and adaptively integrating them based on reliability, effectively handling challenges like low illumination, deformation, and occlusion. CAT [17] addresses shared and specific challenges by using parameter-shared and independent branches, enhancing weak modalities via a guidance module, and forming robust representations for real time. ADRNet [59] identifies key attributes and uses attribute-specific residual branches to build powerful object representations aggregated by an attribute ensemble network. DFNet [31] uses a two-stream structure with non-shared and shared convolution kernels to adaptively weight individual and common features for different sequences. MFGNet [41] enhances RGB-T tracking by adaptively adjusting convolutional kernels for visible and thermal data and using a direction-aware attention mechanism to address occlusion and fast motion. APFNet [45] disentangles the RGBT tracking fusion process via challenge attributes, using five attribute-specific branches and an aggregation module to achieve robust fusion with fewer parameters and reduced dependence on large-scale training data. DMCNet [26] introduces a duality-gated mutual condition network for RGBT tracking, which enhances target representations by using discriminative information from one modality to guide feature learning in another, suppresses noise effects, and employs a resampling strategy to handle sudden camera motion. DiMP-RGBT, based on DiMP [3], explores an end-to-end training framework that evaluates all three fusion strategies (early, middle, and late), finding middle fusion to be the most effective. CMPP uses an inter-modal pattern-propagation process to transfer patterns between modalities. SiamFT [62] was the first to apply the Siamese architecture to RGB-Thermal tracking, proposing an RGB-infrared fusion tracking method based on fully convolutional Siamese Networks, which processes visible and infrared images separately, fuses their features adaptively, and predicts tracking results in real time using a modality weight computation method. SiamIVFN [13] proposes a Siamese infrared and visible light fusion network for RGB-T tracking, containing a complementary-feature-fusion network (CFFN) and a contribution-aggregation network (CAN) to extract both individual and common features. SiamCDA [61] proposes a new RGB-T tracker based on Siamese networks, incorporating a feature pyramid network (FPN) for cross-level information capture, a complementarity-aware multi-modal feature fusion module (CA-MF) for cross-modal information, and a distractor-aware region proposal selection module (DAS) for robustness. DMCNet [26] introduces a duality-gated mutual condition network to leverage discriminative information while minimizing data noise. Recent advances in Parameter-Efficient Fine-Tuning(PEFT) [15, 39, 40, 65] have inspired methods to adapt RGB-modal trackers for multi-modal tracking. Protrack [50] proposes a prompt learning paradigm to harness both large-scale RGB knowledge and complementary information. ViPT [67] introduces a modality-complementary prompter for generating visual prompts in multi-modal tracking. MPLT [28] designs a multi-modal visual information prompter to adaptively generate weights from another modality. These PEFT-based methods are promising, as they enable the use of pure RGB-modal models with fewer parameters, partially addressing the challenge of limited multi-modal data availability. GMMT [35] employs generative models (CGANs and Diffusion Models) to fuse RGB and thermal infrared (TIR) features with random noise, enhancing discriminative feature extraction for robust tracking.

Siam_MMSTC [60], as shown in Fig. 4.9, operates through a structured pipeline comprising six key modules: feature extraction, Multi-modal Transformer Encoder (MMTE), memory pool construction, Quality-aware Transformer Decoder (QATD), target state estimation, and memory update. The feature extraction module employs two-stream Siamese networks, one for RGB and one for thermal infrared inputs, using a modified ResNet-50 backbone. The backbone removes down-sampling operations in the fourth stage and replaces traditional 3 × 3 convolutions with atrous convolutions (rate 2) to enlarge receptive fields, outputting feature maps from the third and fourth stages. For a given RGB-T frame, the RGB and thermal template images and search images are processed to extract unimodal template features ($\mathbf{Z}_3^r, \mathbf{Z}_4^r$ for RGB; $\mathbf{Z}_3^t, \mathbf{Z}_4^t$

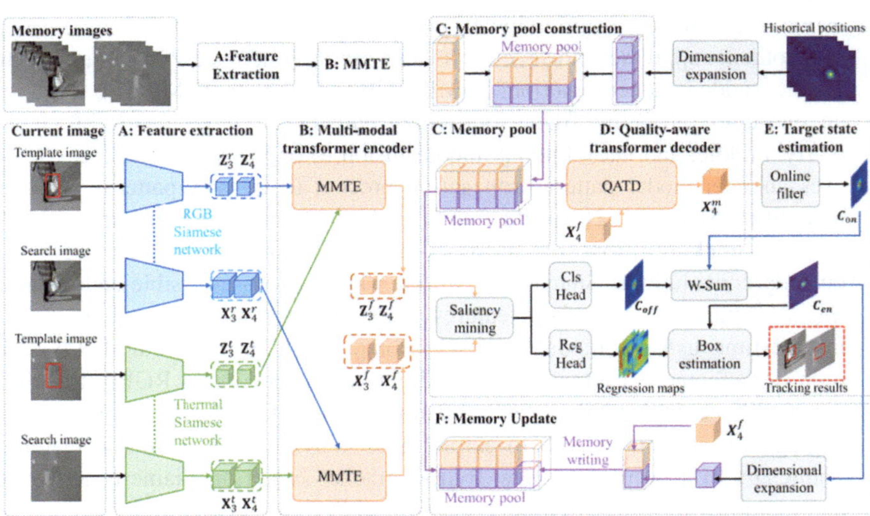

Fig. 4.9 The main architecture diagram of Siam_MMSTC [60]

for thermal) and search features ($\mathbf{X}_3^r$, $\mathbf{X}_4^r$, $\mathbf{X}_3^t$, $\mathbf{X}_4^t$). Template features are cropped from feature maps based on the initial bounding box and pooled using a PrPool layer.

The MMTE module is central to modeling multi-modal spatial contexts and integrating features across modalities. It takes unimodal template and search features as inputs and simultaneously explores intra-modal and cross-modal spatial contexts to produce encoded template and search features. The MMTE employs a Transformer encoder to compute reliable context information by enhancing intra-modal contexts with cross-modal interactions. For RGB and thermal feature maps, the attention mechanism within the Transformer encoder is defined as:

$$A = \text{Softmax}\left(\frac{QK^T}{\sqrt{d_k}}\right)V, \tag{4.55}$$

where Q, K, and V are the query, key, and value matrices derived from input features, and d_k is the dimension of the keys. The MMTE processes feature maps by first concatenating RGB and thermal features along the channel dimension, then applying self-attention to capture intra-modal relationships and cross-attention to embed cross-modal contexts, enhancing the discriminative power of the encoded features.

The memory pool construction module aggregates historical encoded search features and their corresponding tracking results, forming a repository of temporal context information. This memory pool serves as a dynamic knowledge base that captures the evolution of target and background states across frames. The QATD module then leverages this memory pool to propagate temporal context information to the current frame, enhancing the encoded search features. The QATD employs a Transformer decoder that selectively integrates target-related and background-related contexts based on their quality. Target-related contexts, which reflect relationships between targets and backgrounds, are fully utilized to distinguish targets from distractors, while background-related contexts are selectively propagated to mitigate interference. The decoder's attention mechanism is formulated as:

$$F = \text{Softmax}\left(\frac{Q_h K_h^T}{\sqrt{d_k}}\right)V_h, \tag{4.56}$$

where Q_h, K_h, and V_h are derived from historical features in the memory pool, and F represents the enhanced feature output. The QATD outputs decoded search features that incorporate temporal dynamics, improving the tracker's ability to adapt to target state changes.

The target state estimation module combines encoded template features, encoded search features, and decoded search features to predict the target's bounding box in the current frame. It uses a classification head to generate confidence scores and a regression head to estimate bounding box coordinates, employing a loss function that combines weighted focal loss for classification and L1 and generalized IoU losses for regression:

$$L_{\text{track}} = L_{\text{cls}} + \lambda_{\text{iou}} L_{\text{iou}} + \lambda_{L_1} L_1, \quad (4.57)$$

where $\lambda_{\text{iou}} = 2$ and $\lambda_{L_1} = 5$. The memory update module continuously refreshes the memory pool with new tracking results and encoded features, ensuring the temporal context remains relevant. Updates occur selectively, prioritizing high-quality tracking results based on classification scores to avoid incorporating erroneous data.

SDSTrack [11] applies a symmetric framework with parameter-efficient fine-tuning to transfer the feature extraction capability of pre-trained RGB trackers to thermal infrared (TIR) modality, enabling effective fusion of RGB and TIR features. It uses an adapter-based tuning strategy to integrate both modalities without over-relying on either. Additionally, a complementary masked patch distillation approach is introduced to improve robustness under challenging conditions. TATrack [37] employs modality prompts and a spatiotemporal interaction (STI) mechanism to integrate RGB and thermal infrared features, enabling cross-modal interaction across longer time scales. This approach enhances the extraction of discriminative features for robust tracking in complex scenarios.

Specifically, TATrack [37], as shown in Fig. 4.10, builds upon a Vision Transformer (ViT) backbone, extending a baseline model that uses modality prompts to adapt a pre-trained RGB tracker for RGBT tasks. The baseline processes RGB and TIR search regions $I_x^r, I_x^t \in \mathbb{R}^{3\times W_x \times H_x}$ and initial templates $I_z^r, I_z^t \in \mathbb{R}^{3\times W_z \times H_z}$, embedding them into 1D tokens $H_{rx}, H_{tx} \in \mathbb{R}^{N_x \times C}$ and $H_{rz}, H_{tz} \in \mathbb{R}^{N_z \times C}$, where $N_x = H_x W_x / P^2$, $N_z = H_z W_z / P^2$, $P \times P$ is the patch resolution, and C is the token dimension. Position embeddings are added, and token sequences are concatenated as $H_r^0 = [H_{rz}, H_{rx}]$ and $H_t^0 = [H_{tz}, H_{tx}]$. The

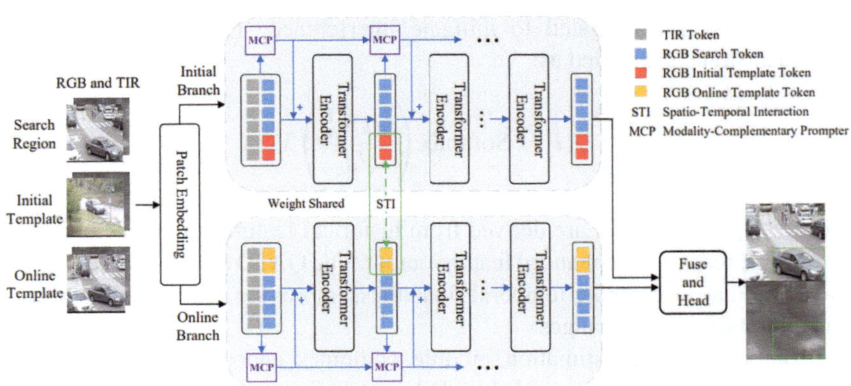

Fig. 4.10 The main architecture diagram of TATrack [37]

ViT backbone, comprising L transformer encoders, extracts features via Multi-head Self-Attention (MSA), LayerNorm (LN), Feed-Forward Network (FFN), and residual connections. The attention function is defined as:

$$A = \text{Softmax}\left(\frac{QK^T}{\sqrt{C}}\right)V = \text{Softmax}\left(\frac{H_r W_q (H_r W_k)^T}{\sqrt{C}}\right)(H_r W_v), \tag{4.58}$$

where Q, K, V are query, key, and value matrices, and W_q, W_k, W_v are projection parameters. The forward propagation is:

$$H_r^l = E^l(H_r^{l-1}), \quad l = 1, 2, \ldots, L. \tag{4.59}$$

A Modality-Complementary Prompter (MCP) generates modality prompts $\mathcal{P}^l = P^l(\mathcal{P}^{l-1}, H_r^{l-1})$, with $\mathcal{P}^0 = H_r^0$, adjusting transformer encoder inputs as $H_r^{l-1} = H_r^{l-1} + \mathcal{P}^l$. MCP reshapes sequences to 2D feature maps, reduces dimensions via 1×1 convolutions, applies spatial fovea operations, and restores dimensions to generate prompts, facilitating cross-modal interaction. A fully convolutional neural network (FCN) head regresses bounding boxes using classification scores, local offsets, and normalized box sizes.

TATrack extends this baseline into a two-branch architecture: an initial branch referencing the initial template and an online branch using a dynamically updated online template. This structure captures temporal information to adapt to target state changes, addressing the baseline's limitation of static template reliance. The initial and online branches process inputs similarly to the baseline, but a Spatiotemporal Interaction (STI) mechanism bridges them, enabling cross-frame propagation of spatiotemporal information. STI concatenates template features Z_i (initial branch) and Z_o (online branch) along the spatial dimension as $Z = \text{Concat}(Z_i, Z_o)$, applying self-attention:

$$F = \text{Softmax}\left(\frac{ZW_q(ZW_k)^T}{\sqrt{C}}\right)(ZW_v), \tag{4.60}$$

followed by:

$$\hat{F} = \text{LN}(F + Z), \quad Z = \text{LN}(\hat{F} + \text{FFN}(\hat{F})). \tag{4.61}$$

STI, inserted before MCP in layers 4, 7, and 10 of the ViT backbone, establishes bi-directional information flow, integrating spatial information from the initial template and temporal information from the online template. This enhances feature extraction and cross-modal interaction, producing robust target representations. The search region sequences from both branches are concatenated, processed by a Conv-BN-ReLU layer to halve channels, and fed to the localization head.

The online template update strategy selects reliable templates based on the maximum target classification score every 50 frames, avoiding poor-quality templates that could mislead tracking. TATrack is trained end to end using a loss function combining weighted focal loss for classification and L1 and generalized IoU losses for regression:

$$L_{\text{track}} = L_{\text{cls}} + \lambda_{\text{iou}} L_{\text{iou}} + \lambda_{L_1} L_1, \tag{4.62}$$

- Late Fusion: SCCF [42] proposes a soft-consistent correlation filter framework. It jointly learns features from RGB and thermal infrared data. The method takes into account both the collaboration and differences between the two modalities. This enables robust visual tracking in adverse and low-light conditions. The framework reduces computational cost using Fast Fourier Transform (FFT). It also introduces a weighted fusion mechanism. This helps improve the accuracy of response map estimation. CSCF [43] creates the final tracking output by merging the response maps from both modalities. JMMAC [58] proposes an RGB-T tracking framework that jointly models appearance and motion cues, using a late fusion method for robust appearance modeling and a tracker switcher to handle unreliable appearance cues by incorporating motion information.

4.3 VOT Method Performance Comparison

4.3.1 RGB-Depth Tracking in VOT

As shown in Table 4.1, the performances of various RGB-D tracking methods are summarized on three widely used benchmark datasets: PTB [32], STC [44], and DepthTrack [44].

At present, most RGB-D approaches rely on CNN-based feature extractors and hand-crafted fusion techniques for combining modalities. However, with the recent success of Vision Transformers in RGB-based tracking, designing effective Transformer-based architectures for multi-modal feature integration is expected to become a key research direction in RGB-D tracking.

Additionally, many existing fusion strategies are non-parametric, such as those based on weighted summation or joint distribution modeling. Developing a learnable and robust fusion network that enables deep interaction between RGB and depth features could significantly enhance tracking performance.

Furthermore, a major limitation of current RGB-D trackers is their inability to meet real-time processing requirements, which is crucial for practical industrial applications. Therefore, improving tracking efficiency remains an important and meaningful research direction.

Table 4.1 Comparison of the performance of different RGB-Depth trackers on RGB-Depth dataset

Fusion	Trackers	Publication	PTB	STC	DepthTrack	
			Success	AUC	Sequence-based F-Score	Frame-based F-Score
Early	PT [34]	ICCV2013	0.733	0.35	–	–
	DS-KCF [8]	BMVC2015	0.690	0.34	0.076	0.075
	3D-T [4]	CVPR2016	0.750	–	–	–
	DS-KCF-Shape [8]	J.RTIP2016	0.719	0.39	0.070	0.071
	CA3DMS [24]	TMM-2018	0.737	0.43	0.224	0.216
Middle	OAPF [30]	CVIU2016	0.731	0.26	–	–
	DLS [1]	ICPR2016	0.739	–	–	–
	DAL [14]	ICPR2019	0.807	0.64	0.429	0.433
	TSDM [63]	ICPR2021	0.792	–	–	–
	DeT [49]	ICCV2021	–	–	0.532	0.502
	ViPT [67]	CVPR2023	–	–	0.596	0.592
	HMAD [48]	ACMMM Asia2023	–	–	0.611	0.598
Late	STC [44]	IEEE TCYB-2018	0.698	0.40	–	–

4.3.2 RGB-Language Tracking in VOT

As presented in Table 4.2, performance results across three benchmark datasets are provided to evaluate and compare RGB-Language-based tracking methods. The evaluation includes setups using only language inputs as well as combinations of language with bounding box annotations, allowing for a more thorough analysis.

Overall, the language modality offers better interpretability for humans and can enhance tracking by providing richer spatial and semantic insights. With recent advances in vision–language foundation models such as CLIP, RGB-Language tracking has the potential to benefit significantly from pre-trained vision–language representations. Unlike Wang et al. [38]. who rely on pre-trained NLP models, leveraging vision–language models could help bridge the gap between visual content and textual semantics through contrastive learning strategies. Furthermore, incorporating linguistic cues opens up possibilities for exploring open-world tracking scenarios, enabling the development of more adaptable and robust tracking systems capable of handling objects beyond conventional categories.

Table 4.2 Comparison of the performance of different RGB-Language trackers

Fusion	Trackers	Publication	LaSOT		OTB-Lang		TNL2K	
			NL	NL+BBox	NL	NL+BBox	NL	NL+BBox
Middle	GTI [51]	TCSVT2020	–	0.47 / 0.47	–	0.73 / 0.58	–	–
	CiteTracker [23]	ICCV2023	–	0.76 / 0.70	–	–	–	0.60 / 0.58
	All-in-One [54]	ACMMM2023	–	0.79 / 0.72	–	0.93 / 0.71	–	0.57 / 0.55
	OVLM [55]	TMM2023	–	0.71 / 0.66	–	–	–	0.67 / 0.63
	LMTrack [27]	ISPDS2023	–	0.78 / 0.72	–	0.93 / 0.75	–	0.59 / 0.57
	MMTrack [64]	TCSVT2023	–	0.76 / 0.70	–	0.82 / 0.71	–	0.59 / 0.59
	SATracker [7]	None2024	–	0.80 / 0.72	–	0.94 / 0.74	–	0.64 / 0.61
	UVLTrack [29]	None2024	0.64 / 0.60	0.79 / 0.71	0.83 / 0.64	0.92 / 0.71	0.61 / 0.58	0.69 / 0.65

Table 4.3 Comparison of the performance of different RGB-LiDAR trackers

Fusion	Trackers	Publication	KITTI				
			Car	Pedestrian	Van	Cyclist	Mean
Middle	F-Siamese Tracker [69]	IROS2020	37.15 / 50.60	16.28 / 32.28	–	47.03 / 77.26	–
	PVTrack [47]	IROS2024	–	73.3 / 86.9	–	–	–
	MATNet [46]	IROS2024	–	76.94 / 88.20	–	87.29 / 90.06	82.11 / 89.13
Late	Alireza et al. [2]	ITSC2016	–	–	–	–	–

4.3.3 RGB-LiDAR Tracking in VOT

The results of RGB-LiDAR trackers are summarized in Table 4.3. Currently, the number of available trackers and datasets is limited, as this area of research is still in its early stages. This emerging field has developed from LiDAR-based tracking, aiming to combine the geometric details from LiDAR with the visual appearance features from RGB data. Future work on RGB-LiDAR VOT is expected to follow the advancements in LiDAR tracking by incorporating RGB information into existing LiDAR-based approaches. An important aspect of this research involves exploring effective fusion strategies, as there is a notable difference between 2D and 3D data formats. Bridging this gap can help align the 2D image plane with the 3D world, enhancing the tracking model's perceptual abilities.

4.3.4 RGB-Thermal Tracking in VOT

Table 4.4 provides an overview of the performance, fusion methods, and inference speed of representative RGB-thermal trackers evaluated on four popular RGB-thermal VOT benchmarks.

In summary, the RGB and TIR modalities offer complementary characteristics, making RGB-T tracking well-suited for handling complex tracking tasks under varying weather conditions, which is crucial for many practical applications. The core of RGB-T tracking lies in effectively extracting and integrating features from both RGB and thermal data, as well as modeling their interactions. Looking ahead, future work should focus on developing more powerful feature extraction techniques and advanced cross-modal relationship modeling, with potential exploration of architectures such as Transformers.

Table 4.4 Comparison of the performance of different RGB-Thermal trackers

Fusion	Trackers	Publication	GTOT		RGBT234		RGBT210		FPS
			PR(%)	SR(%)	PR(%)	SR(%)	PR(%)	SR(%)	
Middle	Cross-modal [18]	ECCV2018	82.7	84.3	–	–	69.4	46.3	8
	DenseFuse [21]	TIP2018	–	66.5	76.4	56.3	–	–	1.5
	FTSNet [19]	NC 2018	85.2	62.6	–	–	–	–	15
	LGMG [20]	TCSVT2018	82.9	65.5	–	–	71.1	46.8	7
	TODA [43]	ICIP2018	84.3	67.7	–	–	–	–	0.349
	FANet [68]	TIV2018	88.5	69.8	76.3	53.2	–	–	1.3
	SiamFT [62]	Access2019	–	–	65.1	44.8	65	44.3	25
	MANet [46]	ICCV2019	89.3	72.4	77.7	53.92	–	–	1.1
	Fast RGBT [53]	NC 2019	77	63.2	–	–	52.8	–	227
	CMPP [36]	CVPR2020	92.58	73.8	82.3	57.4	–	–	1.3
	CAT [17]	ECCV2020	88.9	71.7	80.3	56.1	79.2	53.3	20
	SiamIVFN [13]	None2021	91.5	79.3	81.1	63.2	–	–	147.5
	ADRNet [59]	IJCV2021	90.4	73.9	79.7	63.7	–	–	25
	DFNet [31]	None2021	81.8	71.9	77.2	53.6	–	–	28.6
	MFGNet [41]	TMM2021	88.8	70.7	78.2	53.5	74.9	49.4	3.37
	SiamCDA [61]	TCSVT2021	87.7	73.2	76.0	56.9	–	–	37
	APFNet [45]	AAAI2022	90.5	73.6	82.7	57.9	–	–	–
	DMCNet [26]	TNNLS2022	90.9	73.4	84.0	59.2	79.60	55.4	2.38
	Protrack [50]	ACM MM 2022	–	–	78.6	58.6	–	–	–
	ViPT [67]	CVPR2023	–	–	83.5	61.8	–	–	38.5
	MPLT [28]	None2023	–	–	88.4	65.7	86.20	63.0	22.8
	BAT [5]	None2023	–	–	86.7	64.2	–	–	–
	OneTracker [10]	CVPR2024	–	–	85.7	64.2	–	–	–
	SDSTrack [11]	CVPR2024	–	–	84.8	62.5	–	–	20.86
	TATrack [37]	EAAI2024	–	–	–	–	–	–	30
Late	SCCF [42]	PRCV2018	85	69	–	–	–	–	50
	CSCF [43]	BICS2018	83.3	–	–	–	66.8	37.2	50
	JMMAC [58]	TIP2021	90.1	73.2	79.0	53.2	–	–	26

4.4 Summary and Challenges

Multi-modal target tracking greatly improves tracking performance by combining different types of data. Common combinations include RGB-Depth, RGB-Thermal, RGB-LiDAR, and RGB-Language. These methods use multiple sensors to capture more complete information about the environment. Depending on the fusion stage, these modalities can be categorized into three strategies: early fusion, mid-level fusion, and late fusion. Early fusion combines data from different modalities at the input level for joint feature extraction, fully utilizing all available information from the outset. Mid-level fusion first extracts features independently from each modality and then merges them at the feature level, enriching the information while preserving the unique characteristics of each modality. Late fusion processes each modality independently and integrates the results at the final decision-making stage, ensuring modality independence and maximizing their strengths. By using these fusion strategies, multi-modal target tracking systems can gain a complete understanding of the environment. This leads to better accuracy and stronger performance, even in challenging conditions. Each fusion strategy has its own unique benefits and is best suited for certain types of applications. This variety allows for flexible and efficient tracking solutions across different scenarios. This classification spans the entire process—from data collection to decision-making—and highlights optimization directions for various applications, advancing the development and adoption of multi-modal technologies.

However, multi-modal target tracking still faces numerous challenges, which can be outlined as follows:

- **Feature Complexity and Analytical Difficulties:** The feature spaces of different modalities vary significantly in complexity. For instance, the visual features of RGB images differ greatly from the spatial features of depth data. This can lead to the loss of critical information during feature extraction and fusion, especially in early fusion, where this heterogeneity may exacerbate information loss.
- **Computational Cost and Real-Time Performance Issues:** The high dimensionality and heterogeneity of multi-modal data substantially increase computational complexity, particularly in early fusion, where joint feature extraction may impose a heavy computational burden. This can affect applications with strict real-time requirements, such as target tracking in autonomous driving.
- **Absence of Modalities with High Annotation Dependency:** For example, in RGB-Thermal tracking, thermal data may be missing due to environmental interference. Extreme weather is one possible cause. When this happens, the system may not fuse other modalities well. This can lead to worse tracking performance. In addition, different modalities often depend on each other. If the system relies too much on one modality, performance can get even worse when that modality is not available.
- **Lack of Cross-Domain Transferability:** Current models are typically designed for specific domains. For example, an RGB-Depth fusion model may not work well when directly applied to RGB-Language scenarios. The differences between modalities can cause performance degradation in cross-domain applications. This limits the broader adoption of multi-modal technologies.
- **Data Alignment and Synchronization Issues:** Misalignment of different modalities in time and space, such as timestamp discrepancies between RGB and LiDAR data. It may result in inconsistent fused features, thereby affecting tracking accuracy.
- **Lack of Large-Scale Annotated Multi-modal Datasets:** Building multi-modal datasets requires synchronized annotation across multiple modalities, which is costly and time-consuming. For example, in RGB-Language tracking, language descriptions need to be annotated manually. This process requires a lot of effort. As a result, it becomes difficult to meet the demands of large-scale training. This limitation affects both the model's training and its ability to generalize.

References

1. An N, Zhao X-G, Hou Z-G (2016) Online RGB-D tracking via detection-learning-segmentation. In: 2016 23rd International conference on pattern recognition (ICPR). IEEE, New York, pp 1231–1236
2. Asvadi A, et al (2016) 3D object tracking using RGB and LIDAR data. In: 2016 IEEE 19th International conference on intelligent transportation systems (ITSC). IEEE, New York, pp 1255–1260
3. Bhat G, et al (2019) Learning discriminative model prediction for tracking. In: Proceedings of the IEEE/CVF international conference on computer vision, pp 6182–6191
4. Bibi A, Zhang T, Ghanem B (2016) 3d part-based sparse tracker with automatic synchronization and registration. In: Proceedings of the IEEE conference on computer vision and pattern recognition, pp 1439–1448
5. Cao B, et al (2024) Bi-directional adapter for multimodal tracking. In: Proceedings of the AAAI conference on artificial intelligence, vol. 38(2), pp 927–935
6. Chapelle O (2007) Training a support vector machine in the primal. Neural Comput 19(5):1155–1178
7. Ge J, et al (2023) Beyond visual cues: synchronously exploring target-centric semantics for vision-language tracking. arXiv preprint arXiv:2311.17085
8. Hannuna S, et al (2019) Ds-kcf: a real-time tracker for rgb-d data. J Real-Time Image Proc 16:1439–1458
9. Henriques JF, et al (2014) High-speed tracking with kernelized correlation filters. IEEE Trans Pattern Anal Mach Intell 37(3):583–596
10. Hong L, et al (2024) Onetracker: unifying visual object tracking with foundation models and efficient tuning. In: Proceedings of the IEEE/CVF conference on computer vision and pattern recognition, pp 19079–19091
11. Hou X, et al (2024) Sdstrack: self-distillation symmetric adapter learning for multi-modal visual object tracking. In: Proceedings of the IEEE/CVF conference on computer vision and pattern recognition, pp 26551–26561
12. Ikemura K, et al (2024) Robust depth enhancement via polarization prompt fusion tuning. In: Proceedings of the IEEE/CVF conference on computer vision and pattern recognition, pp 20710–20720
13. Jingchao P, et al (2021) Siamese infrared and visible light fusion network for RGB-T tracking. arXiv preprint arXiv:2103.07302
14. Kart U, et al (2019) Object tracking by reconstruction with view-specific discriminative correlation filters. In: Proceedings of the IEEE/CVF conference on computer vision and pattern recognition, pp 1339–1348
15. Khattak UM, et al (2023) Maple: multi-modal prompt learning. In: Proceedings of the IEEE/CVF conference on computer vision and pattern recognition, pp 19113–19122
16. Li B, et al (2019) Siamrpn++: evolution of Siamese visual tracking with very deep networks. In: Proceedings of the IEEE/CVF conference on computer vision and pattern recognition, pp 4282–4291
17. Li C, et al (2020) Challenge-aware RGBT tracking. In: European Conference on Computer Vision. Springer, Berlin, pp 222–237
18. Li C, et al (2018) Cross-modal ranking with soft consistency and noisy labels for robust RGB-T tracking. In: Proceedings of the European conference on computer vision (ECCV), pp 808–823
19. Li C, et al (2018) Fusing two-stream convolutional neural networks for RGB-T object tracking. Neurocomputing 281:78–85
20. Li C, et al (2018) Learning local-global multi-graph descriptors for RGB- T object tracking. IEEE Trans Circuits Syst Video Technol 29(10):2913–2926

21. Li H, Wu X-J (2018) DenseFuse: a fusion approach to infrared and visible images. IEEE Trans Image Process 28(5):2614–2623
22. Li X, et al (2025) Dynamic Updates for Language Adaptation in Visual-Language Tracking. arXiv preprint arXiv:2503.06621
23. Li X, et al (2023) CiteTracker: correlating image and text for visual tracking. In: Proceedings of the IEEE/CVF international conference on computer vision, pp 9974–9983
24. Liu Y, et al (2018) Context-aware three-dimensional mean-shift with occlusion handling for robust object tracking in RGB-D videos. IEEE Trans Multimedia 21(3):664–677
25. Long Li C, et al (2019) Multi-adapter RGBT tracking. In: Proceedings of the IEEE/CVF international conference on computer vision workshops, pp 0–0
26. Lu A, et al (2022) Duality-gated mutual condition network for RGBT tracking. IEEE Trans Neural Networks Learn Syst 36:4118–4131
27. Lu Q, et al (2023) Natural language guided attention mixer for object tracking. In: 2023 4th International conference on information science, parallel and distributed systems (ISPDS). IEEE, New York, pp 160–164
28. Luo Y, et al (2023) RGB-T tracking via multi-modal mutual prompt learning. arXiv preprint arXiv:2308.16386
29. Ma Y, et al (2024) Unifying visual and vision-language tracking via contrastive learning. In: Proceedings of the AAAI conference on artificial intelligence, vol. 38(5), 4107–4116
30. Meshgi K, et al (2016) An occlusion-aware particle filter tracker to handle complex and persistent occlusions. Comput Vis Image Underst 150:81–94
31. Peng J, Zhao H, Hu Z (2021) Dynamic Fusion Network for RGBT Tracking. arXiv preprint arXiv:2109.07662
32. Qi H, et al (2020) P2b: point-to-box network for 3d object tracking in point clouds. In: Proceedings of the IEEE/CVF conference on computer vision and pattern recognition, pp 6329–6338
33. Shao Y, et al (2024) Context-aware integration of language and visual references for natural language tracking. In: Proceedings of the IEEE/CVF conference on computer vision and pattern recognition, pp 19208–19217
34. Song S, Xiao J (2013) Tracking revisited using RGBD camera: unified benchmark and baselines. In: Proceedings of the IEEE international conference on computer vision, pp 233–240
35. Tang Z, et al (2024) Generative-based fusion mechanism for multi-modal tracking. In: Proceedings of the AAAI conference on artificial intelligence, vol. 38(6), pp 5189–5197
36. Wang C, et al (2020) Cross-modal pattern-propagation for RGB-T tracking. In: Proceedings of the IEEE/CVF conference on computer vision and pattern recognition, pp 7064–7073
37. Wang H, et al (2024) Temporal adaptive RGBT tracking with modality prompt. In: Proceedings of the AAAI conference on artificial intelligence, vol 38(6), pp 5436–5444
38. Wang L, Hui L, Xie J (2021) Facilitating 3D object tracking in point clouds with image semantics and geometry. In: Chinese conference on pattern recognition and computer vision (PRCV). Springer, Berlin, pp 589–601
39. Wang M, et al (2023) Actionclip: adapting language-image pretrained models for video action recognition. IEEE Trans Neural Networks Learn Syst 36:625–637
40. Wang M, et al (2024) M2-clip: a multimodal, multi-task adapting framework for video action recognition. arXiv preprint arXiv:2401.11649
41. Wang X, et al (2021) MFGNet: dynamic modality-aware filter generation for RGB-T tracking. arXiv preprint arXiv:2107.10433
42. Wang Y, Li C, Tang J (2018) Learning soft-consistent correlation filters for RGB-T object tracking. In: Chinese conference on pattern recognition and computer vision (PRCV). Springer, Berlin, pp 295–306

43. Wang Y, et al (2018) Learning collaborative sparse correlation filter for real-time multispectral object tracking. In: International conference on brain inspired cognitive systems. Springer, Berlin, pp 462–472
44. Xiao J, et al (2017) Robust fusion of color and depth data for RGB-D target tracking using adaptive range-invariant depth models and spatio-temporal consistency constraints. IEEE transactions on cybernetics 48(8), 2485–2499
45. Xiao Y, et al (2022) Attribute-based Progressive Fusion Network for RGBT Tracking
46. Xin S, et al (2024) A robotic-centric paradigm for 3D human tracking under complex environments using multi-modal adaptation. In: 2024 IEEE/RSJ international conference on intelligent robots and systems (IROS). IEEE, New York, pp 4934–4940
47. Xin S, et al (2024) Multi-modal 3d human tracking for robots in complex environment with siamese point-video transformer. In: 2024 IEEE international conference on robotics and automation (ICRA). IEEE, New York, pp 337–344
48. Xu B, et al (2023) RGB-D tracking via hierarchical modality aggregation and distribution network. In: Proceedings of the 5th ACM international conference on multimedia in Asia, pp 1–7
49. Yan S, et al (2021) Depthtrack: unveiling the power of rgbd tracking. In: Proceedings of the IEEE/CVF international conference on computer vision, pp 10725–10733
50. Yang J, et al (2022) Prompting for multi-modal tracking. In: Proceedings of the 30th ACM international conference on multimedia, pp 3492–3500
51. Yang Z, et al (2020) Grounding-tracking-integration. IEEE Trans Circuits Syst Video Technol 31(9), 3433–3443
52. Yu C, et al (2024) TF-CLIP: learning text-free CLIP for video-based person re-identification. In: Proceedings of the AAAI conference on artificial intelligence, vol. 38(7), 6764–6772
53. Zhai S, et al (2019) Fast RGB-T tracking via cross-modal correlation filters. Neurocomputing 334:172–181
54. Zhang C, et al (2023) All in One: Exploring Unified Vision-Language Tracking with Multi-Modal Alignment. In: Proceedings of the 31st ACM International Conference on Multimedia, pp 5552–5561
55. Zhang H, et al (2023) One-stream vision-language memory network for object tracking. IEEE Trans Multimedia 26:1720–1730
56. Zhang H, et al (2020) Object tracking in RGB-T videos using modal-aware attention network and competitive learning. Sensors 20(2):393
57. Zhang L, et al (2019) Multi-modal fusion for end-to-end rgb-t tracking. In: Proceedings of the IEEE/CVF international conference on computer vision workshops, pp 0–0
58. Zhang P, et al (2021) Jointly modeling motion and appearance cues for robust RGB-T tracking. IEEE Trans Image Process 30, 3335–3347
59. Zhang P, et al (2021) Learning adaptive attribute-driven representation for real-time rgb-t tracking. Int J Comput Vis 129(9), 2714–2729
60. Zhang T, et al (2024) Exploring multi-modal spatial-temporal contexts for high-performance RGB-T tracking. IEEE Trans Image Process 33:4303–4318
61. Zhang T, et al (2021) SiamCDA: complementarity-and distractor-aware RGB-T tracking based on Siamese network. IEEE Trans Circuits Syst Video Technol 32(3), 1403–1417
62. Zhang X, et al (2019) SiamFT: an RGB-infrared fusion tracking method via fully convolutional siamese networks. IEEE Access 7, 122122–122133
63. Zhao P, et al (2021) Tsdm: tracking by siamrpn++ with a depth-refiner and a mask-generator. In: 2020 25th International conference on pattern recognition (ICPR). IEEE, New York, pp 670–676
64. Zheng Y, et al (2023) Towards unified token learning for vision-language tracking. In: IEEE Trans Circuits Syst Video Technol

65. Zhou K, et al (2022) Conditional prompt learning for vision-language models. In: Proceedings of the IEEE/CVF conference on computer vision and pattern recognition, pp 16816–16825
66. Zhou L, et al (2023) Joint visual grounding and tracking with natural language specification. In: Proceedings of the IEEE/CVF conference on computer vision and pattern recognition, pp 23151–23160
67. Zhu J, et al (2023) Visual prompt multi-modal tracking. In: Proceedings of the IEEE/CVF conference on computer vision and pattern recognition, pp 9516–9526
68. Zhu Y, et al (2020) Quality-aware feature aggregation network for robust RGBT tracking. IEEE Trans Intell Veh 6(1), 121–130
69. Zou H, et al (2020) F-siamese tracker: a frustum-based double siamese network for 3d single object tracking. In: 2020 IEEE/RSJ international conference on intelligent robots and systems (IROS). IEEE, New York, pp 8133–8139

Chapter 5
Datasets and Evaluation Protocols

Abstract This chapter presents a comprehensive review of datasets and evaluation protocols in visual object tracking (VOT). It begins by categorizing datasets based on sensory modality, covering single-modality datasets such as RGB, thermal, LiDAR, and event-based sources, and extending to multi-modal datasets like RGB-depth, RGB-thermal, RGB-LiDAR, and RGB-language. For each type, the chapter outlines representative benchmarks, data characteristics, and application scenarios. It also explores the fusion strategies employed in multi-modal datasets and their impact on performance in complex environments. Following the dataset overview, the chapter introduces standard evaluation protocols used in VOT research. Key performance metrics such as precision, success rate, expected average overlap (EAO), and area under the curve (AUC) are explained, along with evaluation procedures including One Pass Evaluation (OPE), Temporal Robustness Evaluation (TRE), and Spatial Robustness Evaluation (SRE). By summarizing both dataset benchmarks and evaluation methodologies, this chapter provides a structured perspective on the current landscape of VOT research and supports the development of robust and generalizable tracking algorithms.

5.1 VOT Existing Benchmarks

5.1.1 Single-Modal VOT Datasets

5.1.1.1 RGB-Based

OTB The OTB dataset [39] is one of the early and very influential standard benchmarks in the field of visual target tracking research. It was proposed by Yi Wu et al. in 2013, with the intention of only systematically evaluating and comparing the performance of various types of online visual object tracking (VOT) algorithms. The dataset contains 100 fully annotated video sequences with a total frame count of more than 29,000, where each frame provides accurate target bounding box annotations to satisfy a wide range of typical tracking challenge scenarios such as illumination change, occlusion, scale change, fast motion, background interference,

M. Wang et al., *Visual Object Tracking across Modalities*, Advances in Computer Vision and Pattern Recognition, https://doi.org/10.1007/978-981-95-3664-1_5

rotation, and low resolution. OTB labels each sequence with 11 attributes to support detailed performance analysis. Based on these attributes, it constructs challenging data subsets to help researchers analyze differences in algorithm performance under specific problems. In addition, to fully balance the robustness of the algorithms, OTB designed three evaluation modes: OPE initializes with a true frame from the first frame, tracking the whole video in its entirety; TRE divides the video into multiple segments with different starting frames, initializing from different points in time to assess the temporal robustness; SRE perturbs the initialization frame in the first stitch through positional cheapness and scale changes to assess the algorithms' initialization error Sensitivity. Meanwhile, OTB evaluates the tracker using two metrics, accuracy and area under the line (AUC) of the success plot.

TC128 Temple Color-128 (TC128) [23] is the first large-scale rgb-color video dataset constructed specifically to study the role of over-color information in visual target tracking, proposed by Liang et al. [23]. The dataset consists of a total of 128 fully labeled color video sequences, each of which includes 27 object categories covering diverse environments such as indoor and outdoor, motion scenes, and complex environments, and each of which is meticulously labeled with precise target bounding box annotations and the same 11 categories of attributes as the OTB. According to the experimental performance, the color information generally improves the tracking efficiency, especially when dealing with the challenges of rotation, deformation illumination changes, etc. TC128 filled the early gap in color-based tracking evaluation and inspired subsequent research in color modeling and feature fusion.

UAV123 While most of the popular tracking datasets in the past were based on ground-based or scene-specific shots, in UAV123 [32] sequences captured from low-altitude UAVs are included. Low-altitude UAV videos differ significantly from traditional datasets. Targets in UAV123 are generally smaller and more affected by airflow and electromagnetic interference, leading to degraded video quality. The UAV123 dataset contains a total of 123 fully labeled short video sequences with over 110K frames of image data, all of which use upright bounding boxes to accurately define the location and range of the target. Designed for specific scenarios that require long-term tracking of targets, UAV20L contains 20 long videos. The video sequences in UAV20L have longer time spans and more complex target motion trajectories, which greatly advance the development of long-term tracking algorithms.

NFS The NFS dataset [9] is the first benchmark dataset designed specifically for high frame rate target tracking. The dataset includes 100 real-world videos with 380,000 frames captured at 240 FPS. Each frame is labeled with axis-aligned bounding boxes, and each sequence has nine manually annotated visual attributes. There are two versions of the NFS, namely NFS240 and NFS30, NFS240 is the original version where all videos are shot at a high frame rate of 240 FPS with more detailed frame-to-frame variations and virtually no motion blur for evaluating target tracking performance in high frame rate videos. NFS30 is used to analyze the robustness of the tracking algorithms under low frame rate and strong blur conditions by sampling every 8 frames from NFS240 and synthesizing realistic motion blur using Adobe After Effects to simulate the actual shooting effect of a traditional 30 FPS camera. The two complement each other to provide a unified and comprehensive evaluation platform for different tracking algorithms.

VOT The VOT dataset series [1, 10–14, 16, 17, 30, 34] is one of the most representative evaluation platforms in the field of visual target tracking, and has been released annually since 2013, along with the VOT Challenge, which promotes tracking algorithm performance development and standardized evaluation while defining benchmarks for SOTA tracker performance. The most important feature of the VOT dataset series is that it employs a rotating bounding box to annotate the various tracking challenges and better adapt to target attitude changes in the real world. The most important feature of the VOT series is the use of rotating bounding boxes to annotate the tracking challenges and to better adapt to real-world target attitude changes. The earliest VOT2013 focused on short-term tracking performance evaluation, but with the development so far, the VOT series has been gradually expanded to support a variety of tracking task types. As of VOT2022, the benchmark has expanded to include short-term, long-term, real-time, RGB, and RGB-D tracking tasks. This evolution reflects a broader shift from traditional 2D tracking to multi-modal fusion approaches. The challenge held by VOT not only ranks algorithms but also promotes algorithm innovation through public reporting. VOT has become one of the gold standards in the field of target tracking research precisely because of its fine annotation, rigorous evaluation system, and yearly update mechanism.

TrackingNet TrackingNet [33] is one of the largest VOT datasets designed for real-world applications. It includes over 30,000 YouTube videos with 14.4 million annotated frames. Every frame has a visual object horizontal bounding box and dense annotations generated by automatic interpolation. Trackingnet is divided into a training set and a testing set, where the testing set has 511 video sequences covering different object classes and scenarios, and only the GT of the first frame of each testing sequence is provided, which to upload the model's inference results on the testing set in order to see the validation results and rankings. Similar to OTB, trackingnet uses OPE as an evaluation method, where only the GT of the first frame in the test video is used to initialize the tracker, and the whole video after that is tracked only once, without resetting or backtracking.

LaSOT LaSOT [3] is a high-quality, long-sequence, human-intensively labeled large-scale VOT dataset, which provides rich, long-lived, category-balanced, natural-language-enhanced labeled data for visual target tracking. LaSOT has a total of 1,400 videos, of which 1,120 are in the training set, and 280 in the test set, with a total of more than 3,520,000 frames, and an average video length of 2,512 frames, which is sufficient for evaluating most long-term trackers. LaSOT is manually labeled frame by frame with the Bounding Box and the target occlusion state marked on each frame. LaSOT has 70 categories with an equal number of categories to avoid category bias, and each video has a natural language description to support multi-modal tracking.

GOT-10k GOT-10k [7] is a large-scale, high-diversity benchmark dataset specifically designed for pass-by-target tracking tasks, aiming to drive the generalization capability and real-world application performance of tracking algorithms in real-world scenarios. The dataset is systematically sampled from semantic hierarchies extracted from WordNet and contains a total of more than 10,000 video sequences and 1,500,000 high-quality manually labeled bounding boxes, which comprehensively cover a rich variety of object classes and motion patterns, including people, animals, and traffic vehicles. The training set contains 9340 video sequences covering 480 object categories, and the test set contains 420 video sequences covering 83 brand-new object categories. The training and test sets follow the One-Shot Protocol, where the semantic categories are completely isolated, i.e., the test object categories do not appear in the training set at all, thus avoiding the problem of overfitting to specific categories. The average length of each video is at 127 frames, and each frame is equipped with precise annotations, in addition to the target's visibility score and occlusion status. The dataset also explicitly prohibits the use of external data other than the training set for model training to ensure fairness and consistency of the benchmark.

5.1.1.2 Thermal-Based

LTIR The LTIR dataset [6] is a standard dataset designed specifically for short-term VOT tasks and contains 20 thermal imaging video sequences covering six different classes of objects, such as people, animals, and vehicles. Each video sequence has an average frame length of 563 stitches and covers a wide range of indoor and outdoor scenarios and challenges, including changes in target dynamics, temperature variations, blurring, and camera motion. LTIR is not only suitable for tracking studies in the infrared wavelength band but also incorporates visual sequences to support systematic evaluations via the standard VOT evaluation toolkit.

VOT-TIR VOT-TIR 2015 [4] is the first standardized challenge and dataset focused on short-term visual object thermal infrared (TIR) visual tracking. Built on the LTIR dataset [17], it contains diverse data from 7 different organizations and 8 types of sensors. The dataset is labeled manually instead of rotating the bounding box and introduces TIR-exclusive attributes such as dynamic range change and target temperature change. VOT-TIR 2016 [18] is a comprehensive upgrade of the 2015 version. The dataset is expanded to 25 sequences, new long-duration tracking scenarios are added, and thermal crossover attributes are introduced to refine the challenge description. The dataset has become more challenging by enhancing complex scenarios such as dynamic thermal disturbances and thermal contrast changes.

PTB-TIR PTB-TIR [28] is a challenging thermal infrared image dataset for pedestrian tracking. The dataset contains 60 sequences with manual annotations covering scenarios such as night surveillance and complex terrain. The total number of frames is 30128, each frame provides pedestrian axis-aligned rectangular box labeling as well as nine types of challenge attributes, specially designed for TIR's unique design such as Beijing temperature proximity in the target domain. The resolution of the video sequences is 1280 × 720 pixels (720p), which was the highest resolution design at the time the dataset was created. While attributes such as "Thermal Crossing" and "Dynamic Background Disturbance" were introduced in VOT-TIR 2016, other key scene attributes (e.g., extreme weather, target type, motion patterns) were not included. Target types, motion patterns are not systematically categorized, while the single-scene coverage and lack of long cross-scene sequences result in limited applications of the dataset.

MLSSNet The MLSSNet dataset [24] is a large-scale, high-resolution thermal infrared target tracking benchmark dataset containing 500 high-quality thermal infrared video sequences totaling 228,000 frames, all with a resolution of 1920 × 1080 or higher, which is much higher than the traditional TIR dataset. The dataset covers a wide range of environmental conditions, and 83% of the sequences were captured at night or in low-light conditions, which fully reflects the core advantages of thermal imaging in dark/no-light scenarios. MLSSNet uses rotating rectangular frames to accurately annotate the target conveniently and attach fine-grained attribute labels. In addition, the average sequence length of this dataset is 456 frames, covering scenarios such as reappearance after target disappearance and continuous tracking across shots, which is sufficient for long-time tracking tasks.

MMNet The MMNet dataset [27] is a large-scale dataset designed for thermal infrared target tracking, containing 30 categories, more than 1,100 video sequences, 450,000 frames, and 530,000 labeled bounding boxes, with data covering a wide range of viewpoints captured by hand-held devices, in-vehicle cameras, static surveillance, and UAVs, and spanning both day and night as well as complex lighting scenarios. All images are stored in white-hot style and 8-bit depth and are labeled with frame-by-frame bounding boxes and category labels using semi-automated tools.

LSOTB-TIR The LSOTB-TIR dataset [26] is the largest and most comprehensive thermal infrared target tracking benchmark dataset at the time of publication, containing 1,400 sequences, more than 600,000 frames, and more than 730,000 high-quality labeled bounding boxes. The dataset is divided into a training set with 1280 sequences and 47 types of targets, and an evaluation set with 120 sequences and 22 types of challenge targets, covering four types of scenarios: surveillance, UAV, handheld, and vehicle-mounted, and defining 12 challenge attributes (e.g., thermal crossings, interferences, intensity variations, and occlusions), which comprehensively simulate the real complex environment.

ECO-MM ECO-MM [25] contains 1,100 TIR sequences with 450,000 frames and over 30 target categories. The data are collected using more than 30 types of devices—including infrared cameras of varying resolutions, UAVs, and vehicle-mounted sensors—and span four camera perspectives: overlay, panorama, dynamic tracking, and multi-target cross-scene. The core advantage of this dataset lies in the high diversity of hardware and scenarios: multi-devices cover thermal imagers ranging from civilian-grade to professional-grade, and multi-views simulate target motion patterns in complex environments. The design significantly improves the generalization ability of the dataset, which is especially suitable for training and evaluating cross-device adaptive thermal infrared tracking models.

5.1.1.3 LiDAR-Based

KITTI The KITTI dataset [5] is one of the most influential and widely used multi-task benchmarks for vision research in mobile robotics and autonomous driving. It is published by the Karlsruhe Institute of Technology in Germany and the Toyota Technological Institute in Chicago to advance core tasks such as stereo vision, optical flow estimation, visual odometry, and 3D object detection and tracking in real-world environments. For the visual object LiDAR-based tracking task, the KITTI dataset provides 21 training sequences with annotations, widely covering diverse traffic scenarios such as city streets, highways, and villages. To evaluate the performance of the LiDAR-based multi-target tracker, scenes 0–16 of the 21 sequences from KITTI are usually classified as the training set, scenes 7–18 as the validation set, and scenes 19 and 20 as the test set. Each frame in the video sequence simultaneously contains multiple sources of data such as images, 3D point clouds, and high-precision localization information, which has natural time synchronization properties and annotation accuracy. There are four target categories labeled in the dataset: cars, pedestrians, lorries, and bicycles. There is an imbalance of sample distribution in KITTI, and this imbalance of categories also puts higher requirements on the tracking model in terms of weak category generalization ability to avoid performance degradation on weak categories.

It is worth highlighting that KITTI is essentially a highly multi-modal dataset, which contains rich modalities such as synchronously acquired images (binocular grayscale and RGB), IMU+GPS position information, etc., in addition to the LiDAR point cloud. These multi-source heterogeneous data provide important fundamental support for subsequent modal fusion and cross-modal tracking studies.

nuScenes nuScenes [2], developed by Motional, is a multi-modal dataset designed to support research on precise perception and intelligent decision-making in autonomous driving. nuScenes is collected from two cities in Boston and Singapore, with highly diverse and complex traffic scenarios, covering a wide range of environmental conditions, including daytime, nighttime, and sunny. The dataset is collected from two cities, Boston and Singapore, with highly diverse and complex traffic scenarios, covering daytime, nighttime, rain, shine, etc. nuScenes contains a total of 1,000 real urban road driving clips lasting 20 seconds each, which are selected one by one by the team, highlighting the typical scenarios of high-density traffic, complex interactions, and emergencies, which are challenging and representative. nuScenes is divided according to the ratio of 700 training sets, 150 validation sets, and 150 test sets among the 1000 scenarios, which is a reasonable allocation to help ensure the consistency and comparability of the model's training effect, parameter tuning, and the final performance evaluation in the process of developing the model. nuScenes contains a complete set of autonomous driving sensor data, including simultaneous acquisition from 6 cameras, 1 LIDAR, and 5 millimeter-wave radars, resulting in a 360-degree panoramic sensing capability, which is used to evaluate the performance of the model. The acquisition frequency is as high as 20 Hz per second, but the keyframes for annotation are only extracted at 2 Hz, which introduces a large timing difference problem in the visual target tracking task and puts higher demands on the robustness and inference accuracy of the model.

Waymo Open The Waymo Open dataset [36] is a large-scale dataset built specifically for autonomous driving research, integrating high-resolution camera and LiDAR sensor information. The dataset contains a total of 1,150 scenarios, each with a duration of 20 seconds, and is divided into training, validation, and test sets in a ratio of 792/202/150 for systematic model development and evaluation. The Waymo dataset is much larger than traditional datasets such as KITTI, with higher spatial and temporal resolution. More complex road participant interactions and wider urban and suburban coverage make it more challenging in perceptual characters such as target detection and tracking.

5.1.1.4 Event-Based

FE108 FE108 [42] is a large-scale dataset designed for research on fusing event-based and traditional frame-based vision, which contains a total of 108 video sequences, with a total duration of 1.5 hours, covering 21 categories of common targets, and supporting the simultaneous annotation of both frame and event domains. All the data are acquired by the DAVIS346 camera and annotated with high frequency and high precision with the help of the Vicon system, and the sampling frequency of the event stream reaches 240 Hz, which corresponds to the 1/240s time interval of aggregated event frames. The FE108 dataset covers multiple types of real scenarios, such as high dynamics, low illuminance, and fast motion, and it is the most frequently and richly annotated visual object event-frame fusion tracking dataset.

VisEvent VisEvent [38] is a large-scale bimodal dataset developed specifically for event camera and visible light camera target tracking research. The dataset contains 820 pairs of synchronized RGB-Event videos, totaling more than 370,000 frames of images, which cover 17 categories of typical challenging factors such as low-light, high-speed motion and motion blur, and realistically reflecting difficulties faced by the visual perception in the real environment. All the data of VisEvent are acquired synchronously by the event camera, which can output both the frame map and event stream at the same time. It has the advantage of natural spatiotemporal alignment, which largely avoids the error loss caused by post-processing, has high-density and high-quality frame-by-frame annotation, and supports a variety of fusion strategies and benchmark algorithms for evaluation.

5.1.2 Multi-Modal VOT Datasets

5.1.2.1 RGB-Depth

PTB Princeton Tracking Benchmark (PTB) [35] is a standard dataset designed specifically for RGB-D target tracking tasks, proposed by Princeton University in 2013. The dataset consists of 100 indoor RGB-D video sequences captured by a Kinect device with manually labeled 2D bounding boxes for each frame in the video sequences. In total, 95 of the 100 video sequences are not open to the public and are only allowed to be submitted online for evaluation, while the remaining 5 can be accessed for model validation prior to online submission. The PTB dataset has a high degree of diversity, covering human, animal, and rigid-body objects, three types of targets, involving several real-life scenarios such as living rooms, shops, and offices. PTB also refines three types of tracking errors, placing particular emphasis on evaluating occlusion handling.

STC The STC dataset [40] is a high-quality benchmark dataset designed specifically for RGB-D target tracking tasks, first published by Xiao et al. at IEEE T-Cybernetices in 2018, which is designed to compensate for the shortcomings of the PTB dataset in terms of visual diversity, annotation accuracy, and camera motion.

STC contains 36 RGB-D video sequences (about 9000 frames in total), each captured using two ASUS Xtion cameras—one fixed and one moving. This setup simulates viewpoint variations in real applications and ensures that both static and dynamic scenes are represented. The videos were primarily shot in indoor environments and supplemented with a small amount of low-light or nighttime outdoor footage, which was used to examine the algorithm's performance in low-light conditions. To enable a detailed analysis of the tracking algorithm, STC provides detailed attribute annotations in each frame, covering 13 categories of challenge factors. Despite its relatively small size, STC is still regarded as one of the excellent benchmarks for testing the robustness and adaptability of RGB-D tracking algorithms in real and complex scenes due to its fine-grained design, diverse visual distractors, and precise synchronization annotations.

CDTB CDTB [29] is a highly representative and challenging RGB-D tracking dataset. It contains video sequences captured by various RGB-D sensors in both indoor and outdoor scenes under natural lighting, significantly enhancing real-world adaptability. In total, 80 high-quality video sequences are included in CDTB, with a total of more than 100,000 frames, with an average of more than 1,200 frames per sequence, significantly higher than earlier RGB-D datasets, and all video sequences were captured using Kinect v2 in long-time tracking scenes. In order to achieve an in-depth analysis of the algorithm's performance, CDTB provides a fine-grained annotation for each frame, covering 13 visually challenging attributes including occlusion, target disappearance, scale change, fast motion, low-light conditions, etc. There are a very large number of occluded and disappeared target objects in the 102k frames provided by CDTB, and thus, CDTB puts high demands on the algorithm's re-detection and robustness capabilities.

DepthTrack DepthTrack [41] is a large-scale RGB-D target tracking dataset built to advance the development of deep visual tracking algorithms. Captured by a single Intel RealSense415 camera, DepthTrack features a simultaneous output of RGB and depth maps at the same resolution to ensure uniform data quality. DepthTrack contains 200 video sequences with more than 210,000 frames, covering 46 indoor and outdoor scenarios, which greatly enhances the diversity of target types and environments. Each frame is equipped with high-quality bounding box annotations and covers 15 types of challenging attributes, such as occlusion, fast motion, depth change, and camera jitter. Since its release, DepthTrack has become the largest, most detailed, and most widely used public RGB-D tracking dataset. It provides strong support for both training and evaluating multi-modal tracking algorithms.

HAMMER The HAMMER dataset [8] is a high-quality RGB-D polarimetric depth dataset for real scene depth perception tasks, focusing on the study of the differences in the performance of different types of depth sensors in complex environments. The dataset contains a total of 13 diverse real-world scenes with a total of 7200 image frames, the first nine scenes are used for training, and the remaining four scenes are divided into validation and test sets, providing depth sensor measurements, ultrahigh-quality rendered ground-truth depth maps, and polarimetric image data for each frame at the same time. The dataset contains three types of depth sensor measurements: stereo camera depth, d-ToF sensor depth, and iToF sensor depth. The HAMMER dataset provides cross-modal, high-quality, and synchronously aligned depth and polarimetric data, which provides a unified platform for analyzing the perceptual performance of various types of sensors under different scene conditions.

5.1.2.2 RGB-Thermal

GTOT The GTOT dataset [19] is the first large-scale gray-scale-thermal infrared video target tracking benchmark containing 50 pairs of well-aligned gray-scale and thermal infrared video sequences with a total of more than 15.8k frames of detailed annotated data, which are accompanied by frame-by-frame manual annotations with high consistency, including bounding boxes and challenging attributes. The dataset covers a wide range of real-world scenarios and complex challenges such as low-light, occlusion, thermal crossover, and scale variations to support fine-grained evaluation and multi-modal fusion algorithm studies.

RGBT 210 and RGBT234 RGBT210 [21] and RGBT234 [20] are two generations of RGB-T target tracking datasets constructed successively by Li et al. dedicated to advancing the development of multi-modal visual tracking. RGBT210 is the first large-scale and high-quality RGBT dataset containing 210 pairs of aligned video sequences totaling 210,000 frames and providing comprehensive bounding-box annotations with 12 types of challenge attributes.

Based on this, RGBT234, as an upgrade and extension of RGBT210, further enhances the data scale, which is expanded to 234 video pairs with about 234,000 frames. At the same time, more complex environments are introduced, such as hot weather and dynamic camera scenes), and a two-stage high-precision annotation process is adopted to effectively alleviate the inter-modal alignment error problem. RGBT234 not only inherits the design concepts of RGBT210 but also strengthens the flexibility of occlusion-level annotation and modal fusion. RGBT234 is considered one of the most authoritative and challenging benchmarks in the current RGB-T tracking field.

VOT-RGBT The VOT-RGBT dataset is a subset of the VOT Challenge built specifically for the RGB-T short-term target tracking task and is derived directly from the RGBT234 dataset, which consists of two main versions, VOT-RGBT2019 [12] and VOT-RGBT2020 [10]. The data source of VOT-RGBT2019 is derived from 60 pairs of RGB-T video sequences selected from RGBT234 [20], and the rotating bounding box is added to the original bounding box annotation to make it suitable for accurate assessment of attitude irregularities or rotating targets, which focuses on short-term target tracking. VOT-RGBT2020 follows VOT-RGBT2019 in terms of data, with the exception that a new anchor frame mechanism is added to introduce the "reinitialization challenge" to detect the tracker's ability to reposition after multiple losses, and this dataset is more closely aligned to the problem of non-ideal continuous tracking in complex real-world scenarios.

VTUAV VTUAV [43] is a large-scale, high-resolution, multi-attribute RGB-T tracking benchmark released by Zhang et al. [43]. It is specifically designed for dual-modal UAV-based tracking tasks involving visible and thermal infrared imagery. The dataset includes 500 RGB-T video sequences comprising approximately 1,664,549 aligned RGB-T image pairs. The RGB images have a resolution of 1920 $\times$ 1080, while the thermal infrared images are 640 $\times$ 480. All sequences were captured by professional UAVs under diverse weather conditions, including day, night, rain, fog, and wind. VTUAV is divided into a training set of 250 sequences and a test set of another 250, along with an additional set of 100 pixel-level segmentation masks to support video object segmentation tasks. The tracked targets span 13 subclasses and 5 superclasses—pedestrians, vehicles, animals, trucks, and ships—across 15 representative scenes, such as roads, streets, bridges, parks, and shopping centers. The dataset includes approximately 327,000 bounding box annotations, 24,000 segmentation masks, and about 430,000 attribute labels.

Compared with previous RGB-T datasets such as GTOT, RGBT210, RGBT234, and VOT-RGBT, VTUAV significantly improves in data scale, diversity, and application coverage. It supports both short-term and long-term tracking as well as segmentation prediction, making it one of the most comprehensive and versatile benchmarks available for RGB-T tracking research and model evaluation.

5.1.2.3 RGB-LiDAR

For the RGB-LiDAR VOT, the most commonly used dataset in current research is the KITTI dataset [5] presented in Sect. 5.1.1.3.

5.1.2.4 RGB-Language

LaSOT and LaSOText LaSOT [3] is a large-scale, high-quality benchmark for visual object tracking proposed by Fan et al. [3]. It contains 1,400 video sequences totaling over 3.5 million frames, with each video averaging more than 2,500 frames. All frames are manually annotated with bounding boxes and target absence indicators. Targets may disappear and reappear multiple times within a sequence. The dataset is split into two subsets: 1,120 sequences for training and 280 for testing. In addition, LaSOT provides a natural language description for each sequence, enabling joint visual–language multi-modal tracking and supporting direct target localization using textual guidance.

LaSOText [3] is a natural language-augmented subset of LaSOT. It provides one descriptive sentence per sequence, describing the target's appearance, action, and environment. Designed to advance language-guided tracking research, it allows target formulation through linguistic input rather than bounding boxes, facilitating multi-modal fusion approaches. Linguistic cues serve as a global semantic prior in challenging situations such as occlusion, blurring, or background interference, helping visual trackers more accurately identify targets. LaSOText thus plays a key role in bridging vision and language for semantically driven tracking tasks.

OTB-Lang OTB-Lang [22] is a linguistic extension of the classic OTB100 benchmark. Proposed by Li et al. [22], it adds human-written referential descriptions to 99 of the 100 video sequences in OTB100 (one video was excluded due to lack of reliable linguistic annotation). These descriptions enable the target to be specified through language rather than an initial bounding box, thus introducing the paradigm of semantics-driven tracking. The linguistic annotations allow trackers to be initialized based on natural language inputs, expanding the interface between visual tracking and human-centered description.

TNL2K TNL2K [37] is the first large-scale benchmark specifically designed for language-driven visual tracking, proposed by Wang et al. [37]. It contains 2,000 video sequences totaling 1.24 million frames, sourced from YouTube, surveillance systems, and mobile devices, providing rich scene and target diversity. Unlike conventional datasets that rely on bounding box initialization, TNL2K offers one natural language description per video to uniquely identify the target—covering attributes such as category, appearance, shape, and spatial relationships. This helps mitigate ambiguity in target specification. Each frame includes a bounding box and a visibility flag indicating whether the target is present. In total, 17 challenging attributes are annotated, including camera motion, rotation, deformation, blur, occlusion, thermal crossover, model switching, and adversarial scenarios—the last two being introduced for the first time in a visual tracking benchmark.

5.2 VOT Evaluation Protocols

Precision Precision in VOT is the proportion of frames where the error between the center of the tracking result and the center of the real target is within a certain pixel threshold. In short, it is a measure of how accurately the tracker "locates" the center of the target. The formula is defined as follows:

The Euclidean distance between the center of the prediction box and the center of the Ground Truth box is first calculated, usually with a threshold of 20 pixels, i.e., they are considered to be tracked successfully if their Euclidean distance is within 20 pixels:

$$\rho = \sqrt{(x_2 - x_1)^2 + (y_2 - y_1)^2} \tag{5.1}$$

$$\text{Precision@}\theta = \frac{1}{N}\sum_{i=1}^{N} 1(\rho < \theta) \tag{5.2}$$

In the OTB dataset, precision@20 is used as the main metric to evaluate the tracker, and in general, the threshold of precision is modeled on the 20-pixel threshold in the OTB dataset. In addition, in order to comprehensively evaluate the tracker performance, we usually do not just look at a certain threshold, but calculate the precision for each threshold in the range of 1–50 pixels, and draw a precision curve, the meaning of a certain point on the curve: e.g., the precision of 0.84 at 20 pixels means that 84% of the frames have a central prediction error of less than 20 pixels.

Precision is a commonly used metric for evaluating the performance of trackers, but it has some drawbacks: it does not take into account the size of the target, leading to the fact that for small targets even though the prediction frame is far away from the Ground Truth frame, their Euclidean distance is likely to be still within the threshold.

Normalized Precision Normalized Precision is an improved metric proposed to solve the problem of unfair evaluation of traditional Precision (pixel-level centroid error) at different target scales. "Fixed pixel thresholds cannot accommodate different target sizes," so target scale normalization was introduced to make the threshold a "relative threshold." It makes the error assessment between large and small targets more consistent and fair by normalizing the centroid error to a scale relative to the target size.

Considering the scale size of the Ground Truth box, Precision is normalized to obtain Norm. prec, which takes values between [0, 0.5]. That is, the ratio of the Euclidean distance between the prediction box and the center point of the Ground Truth box to the hypotenuse of the Ground Truth box is judged. The specific formula is as follows:

$$\text{NormCLE}_i = \frac{\sqrt{(x_p^i - x_g^i)^2 + (y_p^i - y_g^i)^2}}{\sqrt{w^i \cdot h^i}} \tag{5.3}$$

As with traditional precision, normalized precision plots a curve and takes a specific point as the evaluation metric:

$$\text{Normalized Precision}(\theta) = \frac{1}{N}\sum_{i=1}^{N} 1(\text{NormCLE}_i < \theta) \tag{5.4}$$

Success Rate In the field of VOT, Success Rate is one of the core metrics for evaluating the performance of tracking algorithms and is mainly used to measure the tracker's ability to accurately frame the target throughout the video sequence. Early tracking evaluations used metrics such as mean center error, but when the tracker loses the target, the output position is often random, and the mean error does not correctly reflect the performance. For this reason, Wu et al. introduced an overlap-based success rate metric in the OTB benchmark (OTB-2013), which is used in conjunction with a center-distance-based accuracy metric. Success Rate is derived from the concept of intersection-union-ratio (IoU) in the field of target detection and is able to measure the accuracy of the tracking results in terms of position and scale simultaneously. The success Rate is calculated by calculating the overlap rate between the tracked frame and the true Success Rate counts the number of frames where the IoU exceeds a specific threshold as a percentage of the total number of frames by calculating the overlap between the tracked frame and the real labeled frame. The core idea is that the higher the overlap between the tracking frame and the real frame, the better the tracking result.

$$\text{SR} = \int_0^1 \frac{1}{N}\sum_{i=1}^{N} \mathbb{I}(\text{IoU}_i \geq t)dt \tag{5.5}$$

AO The AO calculates the average of the intersection and concurrency ratio of the predicted frame to the true frame over the entire video sequence of the tracker. Its calculation formula is as follows:

$$\text{AO} = \frac{1}{N}\sum_{t=1}^{N} \text{IoU}_t \tag{5.6}$$

AO can directly reflect the average positioning accuracy of the tracker, and in long videos, serious errors in a single frame have less impact on the overall AO, but the AO fluctuates more in short videos. AO is generally used in the study for a quick comparison of the overall localization ability of different trackers.

EAO EAO is an important evaluation metric for short-term tracking numbers in the field of VOT, especially widely used in the VOT challenge. This metric aims to comprehensively measure the tracker's accuracy and failure recovery during the effective tracking phase, overcoming the limitations of traditional metrics (e.g., Success Rate) in dynamic scenarios. The core idea of EAO is to calculate the average overlap rate of the tracker during consecutive effective tracking periods while considering the recovery interval after tracking failure. It can respond to determinism and robustness but is not directly derived from accuracy and robustness calculations. This is closely related to its evaluation method, which produces multiple subsequences:

- Accuracy is the average IoU on frames before tracking failure, averaged over all subsequences.
- Robustness is the percentage of subsequence frames successfully tracked, and averaged over all subsequences.
- Tracking failure was defined as the overlap between the ground truth and the predicted target position decreasing below 0.1 and not increasing above that value after at least 10 frames. This definition allows for short-time failure recovery in the short-time tracker.

The symbols are defined as follows:

$$\Phi_{N_s} = \frac{1}{N_s} \sum_{i=1:N_s} \Phi_i \tag{5.7}$$

Φ_{N_s} is the expected average overlap, i.e., calculating an average of Φ_i from the first frame to a desired maximum value is the expected average coverage.

$$\hat{\Phi} = \frac{1}{N_{\text{hi}} - N_{\text{lo}}} \sum_{N_s=N_{\text{lo}}:N_{\text{hi}}} \hat{\Phi}_{N_s} \tag{5.8}$$

But the calculation of EAO in VOT is not $N_s = 1 : N_{\max}$ but is $N_s = N_{\text{low}} : N_{\max}$. $\hat{\Phi}$ is the average EAO score for the metric tracker running on $N_s = N_{\text{low}} : N_{\max}$ with expected average overlap.

AUC Area Under the Curve (AUC) is one of the core metrics for evaluating the performance of tracking algorithms, which is mainly used to comprehensively measure the robustness of the tracker under different levels of stringency. AUC provides a more comprehensive performance evaluation by analyzing the Success Plot under different IoU thresholds, avoiding the limitation of a single threshold.

AUC provides a comprehensive reflection of the tracker's performance under all possible IoU thresholds by calculating the area under the Success Plot. The Success Plot takes the IoU threshold as the horizontal axis (0 to 1) and the corresponding success rate as the vertical axis; the larger the AUC value is, the better the comprehensive performance of the tracker is. It is designed to address the evaluation bias that may result from a single threshold, emphasizing the stability of the algorithm under relaxed to strict conditions. Its calculation formula is as follows:

Firstly, a success rate curve is generated to facilitate a series of IoU thresholds, and for each threshold, the success rate is calculated:

$$S(t) = \frac{1}{N} \sum_{i=1}^{N} \mathbb{I}\left(\mathrm{IoU}_i \geq t\right) \tag{5.9}$$

After plotting the success rate curve, AUC is summed using an integral:

$$\mathrm{AUC} = \int_0^1 S(t)dt \tag{5.10}$$

AUC denotes the average performance of the prediction frame overlapping the true frame Chengdu over the entire range from loose to strict, with the following advantages:

- Robust: no bias due to improper selection of a single threshold;
- Comprehensive: incorporating the performance under multiple thresholds, suitable for different accuracy requirements;
- Strong interpretability: high AUC indicates that the predicted frame overlaps with the real frame more stably, and the tracking effect is better.

Accuracy Accuracy is a metric commonly used to measure the consistency of a tracker's predicted target position with the true position, but its exact definition and calculation may vary depending on the dataset or evaluation protocol. Accuracy typically reflects the tracker's localization accuracy within successfully tracked frames, focusing on a measure of how well the tracked frame matches the true frame. In the VOT Challenge, Accuracy is defined as the average overlap (AO) within the effective tracking phase, i.e., the average value of IoUs of successfully tracked frames.

$$\mathrm{AO} = \frac{1}{N_{\mathrm{valid}}} \sum_{t \in \mathrm{valid}}^{N} \mathrm{IoU}_t \tag{5.11}$$

In addition, some studies consider Accuracy as the inverse of the centroid error or similar to the accuracy in classification tasks, but such usage is less common, so we do not present it too much here.

Robustness Robustness measures the ability of a tracker to maintain tracking in complex scenarios, focusing on immunity and recovery. There are the following common definitions:

- Number of failures: the number of times a decision is lost in a video due to IoU=0.
- Average Failure Interval (AFI): the average number of frames between two failures, with a larger AFI indicating greater robustness.
- Robustness Score: a statistic that combines the number of failures and the clip length distribution in the VOT series of challenges.

FPS FPS indicates the number of frames processed per second by the tracker and is one of the most intuitive parameters to measure computational efficiency, which is commonly evaluated in studies by the average FPS and the maximum FPS. The average FPS reports the average processing speed of the entire video sequence, while the maximum FPS reports the peak speed that the tracker can achieve with hardware support, and the maximum FPS responds to the hardware potential, e.g., autopilot requires more than 30 FPS to ensure real-time response.

OPE The OPE method follows a single run of the tracker from start to finish without any restart or intervention behavior. The aim is to evaluate the tracker's ability to sustain the tracking amount in a complete video sequence. The process of OPE is as follows:

1. Initialization: the first frame gives the real frame of the target.
2. Tracking process: subsequent frames rely solely on the algorithm to predict the target position.
3. Result statistics: the average IoU or success rate of the whole sequence is calculated.

$$\text{Success Rate} = \frac{1}{N}\sum_{t=1}^{N}\delta(\text{IoU}_t \geq \tau) \times 100\% \tag{5.12}$$

Because of its simplicity and efficiency, which can reflect the overall capability of the algorithm, it has been adopted as a standard evaluation method by datasets such as OTB [39], UAV123 [9]. The disadvantage of OPE is that it cannot evaluate the failure recovery capability and is sensitive to transient loss.

TRE The TRE method followed initializes the tracker at different starting frames (not the first frame) of the video to test the robustness of the time dimension. The aim is to verify the sensitivity of the algorithm to the initialization time point. The flow of TRE is as follows:

1. Segmented initialization: the video is divided into multiple sub-sequences, each starting from a random intermediate frame (e.g., the 10%, 20%, ..., 90% frames of the original video).
2. Run OPE independently: run OPE once for each sub-sequence.

3. Performance averaging: Calculate the average IoU or success rate of all subsequences.

$$\text{TRE Score} = \frac{1}{K}\sum_{k=1}^{K} \text{Success Rate}_k \tag{5.13}$$

The OTB dataset [39] uses TRE as a complementary evaluation to test the adaptability of the algorithm to changes in the state of the target. TRE can simulate scenarios where a target appears or reappears after being occluded in the target map, but at the same time, the need to run the tracker multiple times leads to its high computational cost.

SRE The SRE method tests the sensitivity of the initialization by applying spatial perturbations (translation, scaling) to the target frame in the initial frame. The aim is to assess the robustness of the algorithm to initial frame labeling errors. Applying perturbations to the target frame in the initial frame to simulate a real scene can test the tolerance of the algorithm to initial errors, but the disadvantage is that it only tests the initialization phase and does not cover the accumulation of errors during the tracking process. The process is as follows:

1. Perturbation generation: 8 types of perturbations are applied to the initial frame (4 direction translations $\pm 10\%$ + 4 scale changes $\pm 10\%$).
2. Run OPE: Run OPE once for each perturbed initial frame.
3. Performance statistics: calculate the average success rate of all perturbations.

$$\text{SRE Score} = \frac{1}{8}\sum_{i=1}^{8} \text{Success Rate}_i \tag{5.14}$$

OPER The OPER method introduces a failure restart mechanism based on OPE to dynamically recover the trace. Its purpose is to comprehensively evaluate accuracy and recovery capability after failures. OPER was adopted in the early VOT Challenge version and later upgraded to a more stringent short-term protocol. The standardized evaluation process is as follows:

1. Failure Determination: A tracking failure is determined when IoU drops to zero.
2. Restart strategy: Immediately reinitialize the tracker at the next frame using the ground truth annotation.
3. Metrics calculation: Calculate the Expected Average Overlap (EAO) by weighting each valid tracking segment's AO with its segment length N_i, as shown below:

$$\text{EAO} = \frac{\sum_i (N_i \cdot \text{AO}_i)}{\sum_i N_i} \tag{5.15}$$

SRER The SRER method tests the long-term robustness of the algorithm under initial error by combining the initial perturbation of SRE with the restart mechanism of OPER. SRER is able to comprehensively test the stability of the initialization and tracking phases, but at the same time, its methodology is extremely complex and requires a large number of computational resources. The starting process is as follows:

1. Initial perturbation: Apply 8 spatial perturbations to the initial bounding box in the first frame.
2. Dynamic tracking: Perform tracking until failure, then immediately restart according to the OPER rule.
3. Combined Scoring: Calculate the average EAO or success rate across all perturbation scenarios.

VOT-ST The core evaluation framework for VOT short-term tracking revolves around a dynamic reset mechanism, which was originally designed to simulate the need for rapid recovery after a target has briefly disappeared or is severely occluded. According to the VOT Challenge technical reports [10, 15], the protocol builds a highly sensitive performance test environment through a strict failure determination and immediate restart strategy. Specifically, when the intersection and concurrency ratio of the tracker's predicted bounding box to the true labeled box in the current frame drops to zero, the system immediately determines that the tracking has failed and reinitializes the tracker with the true label in the next frame. The repetitive execution of this "fail–restart" cycle forces the algorithm to adjust quickly with limited contextual information, thus exposing its robustness in challenging scenarios such as partial occlusion, deformation, or motion blur.

To quantify the comprehensive performance of trackers in short-term tracking tasks, the VOT committee proposed the core metric of Expected Average Overlap (EAO). EAO's assessment philosophy and calculation results have been introduced previously. This metric has become the gold standard in the post-VOT2016 challenge, as it captures a balanced relationship between tracking accuracy (AO) and failure frequency (Robustness) at the same time. Studies have shown that EAo correlates significantly better with subjective human assessments than traditional Success Rate (SR) and is particularly sensitive to algorithmic stability when dealing with fast motion or sudden changes in lighting.

VOT-LT The proposal of the long-term tracking evaluation protocol stems from the realistic needs of practical applications where the target may be out of the field of view for a long period of time or experience multiple occlusions. Unlike the immediate reset of VOT-ST, VOT-LT adopts a loose failure determination strategy: the system determines tracking failure only when the IoU of the tracker is below the threshold θ (e.g., $\theta = 0.1$) for a consecutive T-frames (usually T=10). Thereafter, the algorithm needs to perform a global search over the whole frame range instead of relying on local context reinitialization. This design forces the tracker to integrate a re-detection module to cope with the challenge of reappearance of the target after a long period of disappearance. The evaluation system of VOT-LT is based on the

F-score as the core metric, which is defined as the reconciled average of precision and recall:

$$F_\beta = \left(1 + \beta^2\right) \frac{\mathrm{Pr} \cdot \mathrm{Re}}{\beta^2 \cdot \mathrm{Pr} + \mathrm{Re}} \tag{5.16}$$

where β is a weighting parameter (usually set to 1), Pr denotes the ratio of the number of successfully detected frames to the total number of detected frames, and Re is the ratio of the number of successfully detected frames to the number of frames actually present in the target. In addition, tracking length is defined as the percentage of frames successfully tracked by the algorithm during the period when the target is visible, which is used to complement the assessment of the algorithm's continuous tracking capability. Notably, VOT-LT requires that the algorithm does not rely on real annotations during the global search phase and resumes tracking entirely through autonomous detection, which poses a serious challenge to traditional trackers based on discriminative models [31].

References

1. Alatan, AA (2017) The Visual Object Tracking VOT2017 Challenge Results
2. Caesar H, et al (2020) nuscenes: a multimodal dataset for autonomous driving. In: Proceedings of the IEEE/CVF conference on computer vision and pattern recognition, pp 11621–11631
3. Fan H et al (2019) Lasot: a high-quality benchmark for large-scale single object tracking. In: Proceedings of the IEEE/CVF conference on computer vision and pattern recognition, pp 5374–5383
4. Felsberg M et al (2015) The thermal infrared visual object tracking VOTTIR2015 challenge results. In: Proceedings of the ieee international conference on computer vision workshops, pp 76–88
5. Geiger A, Lenz P, Urtasun R (2012) Are we ready for autonomous driving? the kitti vision benchmark suite. In: 2012 IEEE conference on computer vision and pattern recognition. IEEE, New York, pp 3354–3361
6. Gundogdu E, et al (2016) Evaluation of feature channels for correlation-filter-based visual object tracking in infrared spectrum. In: Proceedings of the IEEE conference on computer vision and pattern recognition workshops, pp 24–32
7. Huang L, Zhao X, Huang K (2019) Got-10k: a large high-diversity benchmark for generic object tracking in the wild. IEEE Trans Pattern Anal Mach Intell 43(5):1562–1577
8. Jung HJ, et al (2022) Is my Depth Ground-Truth Good Enough? HAMMER–Highly Accurate Multi-Modal Dataset for DEnse 3D Scene Regression. arXiv preprint arXiv:2205.04565
9. Hamed Kiani Galoogahi et al (2017) Need for speed: a benchmark for higher frame rate object tracking. In: Proceedings of the IEEE international conference on computer vision, pp 1125–1134
10. Kristan M, et al (2020) The eighth visual object tracking VOT2020 challenge results. In: Proceedings of the Computer Vision–ECCV 2020 Workshops: Glasgow, UK, August 23–28, 2020, Part V 16. Springer, Berlin, pp 547–601
11. Kristan M, et al (2021) The ninth visual object tracking vot2021 challenge results. In: Proceedings of the IEEE/CVF international conference on computer vision, pp 2711–2738
12. Kristan M, et al (2019) The seventh visual object tracking VOT2019 challenge results. In: Proceedings of the IEEE/CVF international conference on computer vision workshops, pp 0–0

13. Kristan M, et al (2018) The sixth visual object tracking vot2018 challenge results. In: Proceedings of the European conference on computer vision (ECCV) workshops, pp 0–0
14. Kristan M et al (2022) The tenth visual object tracking vot2022 challenge results. In: European Conference on Computer Vision. Springer, Berlin, pp 431–460
15. Kristan M, et al (2015) The visual object tracking VOT2015 challenge results. In: Proceedings of the IEEE international conference on computer vision (ICCV) workshops
16. Kristan M, et al (2015) The visual object tracking vot2015 challenge results. In: Proceedings of the IEEE international conference on computer vision workshops, pp 1–23
17. Kristan M, et al (2014) The vot2013 challenge: overview and additional results
18. Lebeda K, Hadfield SJ, Bowden R, et al (2016) The thermal infrared visual object tracking VOT-TIR2016 challenge result. In: Proceedings, European Conference on Computer Vision (ECCV) workshops
19. Li C et al (2016) Learning collaborative sparse representation for grayscale-thermal tracking. IEEE Trans Image Process 25(12):5743–5756
20. Li C et al (2019) RGB-T object tracking: benchmark and baseline. Pattern Recogn 96:106977
21. Li C, et al (2017) Weighted sparse representation regularized graph learning for RGB-T object tracking. In: Proceedings of the 25th ACM international conference on Multimedia, pp 1856–1864
22. Li Z, et al (2017) Tracking by natural language specification. In: *Proceedings of the IEEE conference on computer vision and pattern recognition*, pp 6495–6503
23. Liang P, Blasch E, Ling H (2015) Encoding color information for visual tracking: Algorithms and benchmark. IEEE Trans Image Process 24(12):5630–5644
24. Liu Q, et al (2020) Learning deep multi-level similarity for thermal infrared object tracking. IEEE Trans Multimedia 23:2114–2126
25. Liu Q, et al (2022) Learning dual-level deep representation for thermal infrared tracking. IEEE Trans Multimedia 25:1269–1281
26. Liu Q, et al (2020) LSOTB-TIR: a large-scale high-diversity thermal infrared object tracking benchmark. In Proceedings of the 28th ACM international conference on multimedia, pp 3847–3856
27. Liu Q, et al (2020) Multi-task driven feature models for thermal infrared tracking. In: Proceedings of the AAAI conference on artificial intelligence, vol. 34, pp 11604–11611
28. Liu Q, et al (2019) PTB-TIR: a thermal infrared pedestrian tracking benchmark. IEEE Trans Multimedia 22(3), 666–675
29. Lukezic A, et al (2019) Cdtb: a color and depth visual object tracking dataset and benchmark. In: Proceedings of the IEEE/CVF international conference on computer vision, 10013–10022
30. Matas J, et al (2015) The visual object tracking VOT2014 challenge results. In: Proceedings of the Computer Vision-ECCV 2014 Workshops: Zurich, Switzerland, September 6–7 and 12, 2014, Part II, vol. 8926. Springer, Berlin, p 191
31. Mayer RE (2020) Searching for the role of emotions in e-learning. Learn Instr 70:101213
32. Mueller M, Smith N, Ghanem B (2016) A benchmark and simulator for UAV tracking. In: Computer Vision–ECCV 2016: 14th European Conference, Amsterdam, The Netherlands, October 11–14, 2016, Proceedings, Part I 14. Springer, Berlin, pp 445–461
33. Muller M, et al (2018) Trackingnet: a large-scale dataset and benchmark for object tracking in the wild. In: Proceedings of the European conference on computer vision (ECCV), pp 300–317
34. Roffo G, Melzi S, et al (2016) The visual object tracking VOT2016 challenge results. In: Proceedings of the Computer Vision–ECCV 2016 Workshops: Amsterdam, The Netherlands, October 8–10 and 15–16, 2016, Part II. Springer International Publishing, Berlin, pp 777–823
35. Song S, Xiao J (2013) Tracking revisited using RGBD camera: unified benchmark and baselines. In: Proceedings of the IEEE international conference on computer vision, pp 233–240
36. Sun P, et al (2020) Scalability in perception for autonomous driving: Waymo open dataset. In: Proceedings of the IEEE/CVF conference on computer vision and pattern recognition, pp 2446–2454

37. Wang X, et al (2021) Towards more flexible and accurate object tracking with natural language: algorithms and benchmark. In: Proceedings of the IEEE/CVF conference on computer vision and pattern recognition, pp 13763–13773
38. Wang X, et al (2023) Visevent: reliable object tracking via collaboration of frame and event flows. IEEE Trans Cybern 54(3), 1997–2010
39. Wu Y, Lim J, Yang M-H (2013) Online object tracking: a benchmark. In: Proceedings of the IEEE conference on computer vision and pattern recognition, pp 2411–2418
40. Xiao J, et al (2017) Robust fusion of color and depth data for RGB-D target tracking using adaptive range-invariant depth models and spatio-temporal consistency constraints. IEEE Trans Cybern 48(8):2485–2499
41. Yan S, et al (2021) Depthtrack: unveiling the power of rgbd tracking. In: Proceedings of the IEEE/CVF international conference on computer vision, pp 10725–10733
42. Zhang J, et al (2021) Object tracking by jointly exploiting frame and event domain. In: Proceedings of the IEEE/CVF international conference on computer vision, pp 13043–13052
43. Zhang P, et al (2022) Visible-thermal UAV tracking: a large-scale benchmark and new baseline. In: Proceedings of the IEEE/CVF conference on computer vision and pattern recognition, pp 8886–8895

Chapter 6
Future Prospects

Abstract This chapter explores the future directions of visual object tracking (VOT), emphasizing emerging techniques that address key research challenges. It highlights the importance of pre-trained large-scale models and the shift toward parameter-efficient fine-tuning (PEFT) for effective model transfer and adaptation. Long-term tracking is discussed in the context of real-world scenarios, focusing on re-detection mechanisms and semantic understanding for robust performance in dynamic environments. The chapter also examines lightweight model design for deployment in resource-constrained settings, unsupervised tracking approaches that leverage unlabeled data to reduce annotation costs, and few-shot tracking techniques that enable fast adaptation with minimal samples. Together, these topics reflect the ongoing evolution of VOT toward greater generalization, efficiency, and adaptability. The chapter is structured around these five major themes, each addressing both the current challenges and the promising advancements driving the field forward.

6.1 Pretrained Large-Scale Model Transfer

With the rapid development of available datasets, model architectures, and training algorithms, a large number of visual base models have been developed. In particular, transformer-based pre-trained visual models (PVMs) have demonstrated excellent performance in a variety of computer tasks. Pre-trained visual macro models learned from massive data have strong generalization capabilities. They capture cross-modal semantic associations (e.g., image-text correspondences) and provide high-quality initial features for downstream tasks. Meanwhile, how to effectively fine-tune PVMs to learn downstream tasks has become a hot issue, as well as how to improve the accuracy and efficiency of downstream tasks by enabling the inheritance of the pre-trained large model's robust understanding of complex scenes. Although traditional full fine-tuning is effective, it requires a large amount of computational and memory resources. This is an almost unacceptable cost for large-scale models with billions or even trillions of parameters. Furthermore, the fact that separate model weights

M. Wang et al., *Visual Object Tracking across Modalities*, Advances in Computer Vision and Pattern Recognition, https://doi.org/10.1007/978-981-95-3664-1_6

need to be maintained for different downstream task authors becomes even more impractical as the number of tasks increases.

Parameter-efficient fine-tuning, originally proposed in NLP, is a promising solution to the above problems. Parameter-efficient fine-tuning aims to overcome the above challenges by updating the least number of parameters while achieving comparable or better performance than full fine-tuning whenever possible. According to recent advances, it has been shown that large pre-trained models trained on rich data have strong generalization capabilities and most of the parameters in PVM can be shared for new tasks. In summary, the PEFT approach can effectively reduce the number of learnable parameters and can protect the a priori knowledge learned from large-scale data in PVMs while effectively adapting to new tasks.

PEFT [22] has become a very popular direction in the past few years, being used in a wide range of tasks from NLP to computer vision. There are also a growing number of approaches in VOT that use effective transfer learning techniques to migrate pre-trained visual models to tracking tasks, such as CPIPTrack [25], TATrack [5], and 3DPT. CPIPTrack directly uses pre-trained CLIP as a bimodal feature extractor and freezes all parameters. This avoids the need for large-scale computational resources and training data, as well as the loss of prior knowledge in traditional full-parameter fine-tuning. Only three types of lightweight learnable cues are added to the CLIP coding layer, through which a small number of learnable cues, each layer of the CLIP encoder can dynamically adjust the cross-modal relationship between visual and verbal, and at the same time, through a small amount of fine-tuning of the downstream task data, the CLIP can be migrated from the general "image-text alignment" capability to "target localization" capability, and then to "target localization." TATrack [5] proposes a modality-complementary cue generator that enables cross-modal interaction between RGB and TIR modalities by embedding them in a ViT with frozen parameters in order to adjust their input features. Previous multi-modal work has taken the form of "pre-trained RGB baseline + multi-modal data fine-tuning," which, however, leads to the problem of knowledge forgetting. Nowadays, PEFT has been shown to be a method that can make good use of the knowledge of pre-trained models at the cost of introducing only a few trainable parameters.

In summary, as pre-trained large-scale model transfer technology continues to evolve, PEFT is gradually transitioning from coarse-grained knowledge transfer to more fine-grained and dynamic adaptation mechanisms. In the field of VOT, compared to earlier methods that achieved representation transfer by freezing the backbone network, recent approaches prefer to utilize efficient parameter fine-tuning strategies. This allows them to retain the advantages of pre-trained models while achieving deep integration and customized optimization of cross-modal knowledge. In the future, PEFT is expected to further enhance the generalization capabilities and task adaptability of multi-modal visual modeling while maintaining computational efficiency.

6.2 Long-Term VOT

Visual target tracking is increasingly focusing on large-scale long-term tracking due to its great potential in real-world applications. Existing target tracking algorithms are often tested and evaluated on videos with an average length of fewer than 1000 frames, and even recently proposed datasets such as LaSOT [3] and TrackingNet [11] contain long-duration videos that are no longer than 3 minutes in length, which is a far cry from the videos used in real-world applications. The main difference between a long-term tracker and a short-term tracker is that the former has to deal with the frequent disappearance and reappearance of the target. Long-term tracking sequences typically last hundreds or thousands of frames. They often involve challenges such as appearance changes, long disappearances, and intra-class interference. For long-term tracking, once the tracker loses the target, subsequent video sequences often become ineffective, which makes long-term target tracking more practical. Therefore, a long-term tracker should have the ability to effectively re-detect the target and be able to distinguish the target from similar interferers.

Recent VOT methods for long-term tracking include SPLT [23], LTMU [2], SiamRCNN [20], DMTrack [1], etc., SPLt utilizes the SiamPRN structure for tracking and re-detection, report this when the target is lost for a long time, it can be relocated to the target through global search. LTMU integrates a starve of the on-line local tracker, verifier, and intermediate updates, which is used by the verifier to try to evaluate the tracking confidence and determine whether the target is lost or not, and proposes an intermediate updater to determine whether the model is updated or not, as a way to avoid model bias due to error accumulation. SiamRCNN combines the efficient feature matching of the Siamese network and the accurate region detection capability of Faster R-CNN to improve the robustness of tracking through dynamic template updating and re-detection mechanism. DMTrack proposes a solution based on dynamic convolution. While the traditional Siamese network lacks flexibility by fusing template and search region features through a fixed structure, DMTrack dynamically generates convolution kernel parameters based on the template features through a dynamic convolution controller and embeds them into the detection cast of the search region, enabling the model to adaptively fuse target information. Despite the progress made so far, the field of long-duration target tracking still has great challenges that require further research and exploration. In addition, in the long-duration target tracking task of complex dynamic scenes, combining linguistic descriptions to solve the target re-identification problem and injecting a priori knowledge through linguistic descriptions are promising to significantly improve the robustness of target re-identification in complex scenes. In the future, with the development of the multi-modal large model and efficient fine-tuning technology, this direction is expected to promote the evolution of visual tracking from "passive response" to "active semantic understanding," and to achieve the low-cost, high-precision continuous perception of dynamic targets in the field of intelligent driving.

6.3 Lightweight VOT Models

In visual target tracking tasks, the backbone network, as a core component of target feature extraction, directly affects the model's ability to characterize the target. In early research, deep networks represented by AlexNet [7], VGGNet-M [17], VGG-16/19, and ResNet series were widely used. These networks significantly improved the feature representation ability by increasing the network depth and channel width, but at the same time led to the problem of computational surge and memory usage. For example, ResNet-50 has up to 25 million parameters, which makes it difficult to be directly deployed in scenarios with limited computational resources, such as mobile devices, drones, or industrial robots. Since 2021, Transformer-based architectures have gradually replaced traditional CNNs in object tracking, emerging as the next-generation backbone networks with superior performance. Notable works include Vision Transformer (ViT), Swin Transformer, DeiT, and architectures specifically designed for tracking tasks such as TransT, STARK, MixFormer, and SimTrack. These methods excel at modelling long-range dependencies and contextual relationships through self-attention mechanisms, significantly improving tracking accuracy. However, their model complexity and inference latency present new challenges. As a result, lightweight design has become a research focus: how to reduce model parameter counts and computational complexity while maintaining tracking performance and build efficient tracking systems deployable on edge devices. Common optimization techniques include lightweight Transformer modules (e.g., MobileViT, EfficientFormer), model pruning, distillation, small model structure redesign, and dynamic inference. These methods have already been applied in practical scenarios such as mobile augmented reality, intelligent security, and medical imaging. Future research should seek a better balance between performance, efficiency, and deployment friendliness and further explore new paradigms such as cross-modal information fusion, self-supervised or semi-supervised learning, to promote the widespread application of lightweight object tracking technology in complex, dynamic, and resource-constrained environments.

6.4 Unsupervised VOT

Unsupervised learning is an important paradigm in machine learning, which does not rely on manually labeled data labels, but rather models the data by mining the intrinsic structure, distribution, and feature relationships of the data itself. In unsupervised learning, models can be learned by the similarity between data, clustering properties, or data distribution patterns, and the method has significant advantages in handling large-scale data and reducing the labeling cost.

Therefore, in recent years, unsupervised learning tracking methods have gradually emerged in VOT research, and unsupervised learning has been introduced in the field of VOT mainly due to the following considerations: the high cost of video annotation and the need for a large amount of data to support DNN-based methods, which has been experimentally verified [9] (e.g., the application of SiamRPN algorithm on the large-scale database Youtube-BB [14]), which means that a larger amount of training data can make the target tracking algorithm obtain better accuracy to a certain extent. In recent years, video resources—such as online videos and surveillance footage—have become extremely abundant, providing large amounts of unlabeled data. However, how to fully utilize these large-scale unlabelled datasets remains a major challenge. Frame-by-frame annotation requires significant human and storage resources. Moreover, manually labeling bounding boxes in every frame limits the model's generalization ability. In contrast, unsupervised learning can leverage these unlabeled resources to train models, significantly reducing the reliance on costly labeled datasets.

Nevertheless, unsupervised tracking still has a great deal of room for development. In traditional supervised target tracking, models are often trained with precise bounding box annotations, which allows the predictions at each step to be directly compared with real labels, thus providing a high degree of controllability and interpretability. However, in unsupervised target tracking, the training data used are unlabeled, which means that the training process does not have the "ground truth" as knowledge, and the model may learn different strategies from human intuition, which may lead to potential error accumulation and drift. Wang et al. realized this and first proposed the UDT algorithm [21] for deep tracking with unsupervised learning. The UDT algorithm uses the consistency between the target forward-tracking prediction and the resultant backtracking response map to compute the network loss and realizes both forward and backpropagation of the network. Regression branch, in which pseudo-labels are difficult to support fine supervisory responsibility, directly leads to the classification branch only roughly locating the target without sensing the target's size change. ULAST [16] effectively establishes the optimization path of the regression branch by means of the micro-maskable mechanism, time-series consistency training, and dynamic pseudo-label weighting, which improves the ability of the supervisable tracking method to adapt to the target's scale and precise location. When the training data are unlabeled, the unsupervised learning methods provide more possibilities for practical applications.

In addition, unsupervised learning methods are also highly competitive in multi-modal VOT, which is caused by the fact that multi-modal VOT usually lacks perfect benchmarks.

6.5 Few-Shot VOT

Few-shot tracking is a cutting-edge technique that deeply integrates few-shot learning with visual target tracking, and its core goal is to achieve continuous accurate target localization in complex scenes through a very small number of annotated samples (e.g., single-frame target initial frames). Traditional tracking methods usually rely on fixed end-shift matching or deep features trained offline, which make it difficult to cope with drastic changes in target appearance and similar interfering objects in the background. As an important research dimension in few-shot tracking research, few-shot learning has formed a systematic theoretical system in the field of few-shot classification, and early research focused on generative models to break through the limitations of sample scarcity through data distribution inference; while in recent years, discriminative modeling has gradually become mainstream, and its breakthroughs originate from the introduction of meta-learning frameworks, where the meta-learning mechanism extracts generic "fast adaptation capabilities" from a large number of tasks in the offline phase, and adapts to target changes with only a small number of historical samples by dynamically optimizing the classifier weights in the online tracking phase. The meta-learning strategy helps the model "learn to track." More importantly, it enables the model to quickly adapt to new target states. Classical three-branch classification of meta-learning is as follows [8, 24]: the parameter-based [4, 13] optimization approach achieves model weight updating on sparse datasets through gradient descent; the metric-based [6, 18, 19] approach constructs a uniform feature projection space so that embeddings of similar samples have spatial proximity, and the model-based approach [10, 12, 15] designs parameterized predictors to directly generate parameter configurations for the model in a feed-forward manner.

Overall, few-shot learning provides solid support for target tracking under conditions of sample scarcity by building stronger generalization capabilities and rapid adaptation mechanisms. With the continuous evolution of modeling strategies, few-shot tracking is expected to achieve more efficient and robust target perception and positioning in complex environments.

References

1. Abdelaziz O (2024) Beyond traditional single object tracking. PhD thesis. University of British Columbia, Columbia
2. Dai K et al (2020) High-performance long-term tracking with meta-updater. In: Proceedings of the IEEE/CVF conference on computer vision and pattern recognition, pp 6298–6307
3. Fan H et al (2019) Lasot: a high-quality benchmark for large-scale single object tracking. In: Proceedings of the IEEE/CVF conference on computer vision and pattern recognition, pp 5374–5383
4. Finn C, Abbeel P, Levine S (2017) Model-agnostic meta-learning for fast adaptation of deep networks. In: International conference on machine learning. PMLR, New York, pp 1126–1135
5. Huang K et al (2024) TATrack: target-aware transformer for object tracking. In: Engineering applications of artificial intelligence, vol 127, p 107304
6. Koch G, Zemel R, Salakhutdinov R, et al (2015) Siamese neural networks for one-shot image recognition. In: ICML deep learning workshop, vol. 2(1). Lille, pp 1–30
7. Krizhevsky A, Sutskever I, Hinton GE (2012) Imagenet classification with deep convolutional neural networks. In: Advances in neural information processing systems, p 25
8. Lee Y, Choi S (2018) Gradient-based meta-learning with learned layerwise metric and subspace. In: International conference on machine learning. PMLR , New York, pp 2927–2936
9. Li B et al (2018) High performance visual tracking with Siamese region proposal network. In: Proceedings of the IEEE conference on computer vision and pattern recognition, pp 8971–8980

10. Mishra N et al (2017) A simple neural attentive meta-learner. In: arXiv preprint arXiv:1707.03141
11. Muller M et al (2018) Trackingnet: a large-scale dataset and benchmark for object tracking in the wild. In: Proceedings of the European conference on computer vision (ECCV), pp 300–317
12. Munkhdalai T, Yu H (2017) Meta networks. In: International conference on machine learning. PMLR, New York, pp 2554–2563
13. Ravi S, Larochelle H (2017) Optimization as a model for few-shot learning. In: International conference on learning representations
14. Real E et al (2017) Youtube-boundingboxes: a large high-precision human-annotated data set for object detection in video. In: Proceedings of the IEEE conference on computer vision and pattern recognition, pp 5296–5305
15. Santoro A et al (2016) Meta-learning with memory-augmented neural networks. In: International conference on machine learning. PMLR, New York, pp 1842–1850
16. Shen Q et al (2022) Unsupervised learning of accurate Siamese tracking. In: Proceedings of the IEEE/CVF conference on computer vision and pattern recognition, pp 8101–8110
17. Simonyan K, Zisserman A (2014) Very deep convolutional networks for large-scale image recognition. arXiv preprint arXiv:1409.1556
18. Snell J, Swersky K, Zemel R (2017) Prototypical networks for few-shot learning. In: Advances in neural information processing systems, vol 30
19. Vinyals O et al (2016) Matching networks for one shot learning. In: Advances in neural information processing systems, vol 29
20. Voigtlaender P et al (2020) Siam r-cnn: visual tracking by re-detection. In: Proceedings of the IEEE/CVF conference on computer vision and pattern recognition, pp 6578–6588
21. Wang N et al (2019) Unsupervised deep tracking. In: Proceedings of the IEEE/CVF conference on computer vision and pattern recognition, pp 1308–1317
22. Xin Y et al (2024) Parameter-efficient fine-tuning for pre-trained vision models: a survey. arXiv preprint arXiv:2402.02242
23. Yan B et al (2019) 'skimming-perusal'tracking: a framework for real-time and robust long-term tracking. In: Proceedings of the IEEE/CVF international conference on computer vision, pp 2385–2393
24. Yao H et al (2020) Automated relational meta-learning. In: arXiv preprint arXiv:2001.00745
25. Zhu H et al (2024) Vision-language tracking with CLIP and interactive prompt learning. In: IEEE transactions on intelligent transportation systems

MIX
Papier aus verantwortungsvollen Quellen
Paper from responsible sources
FSC® C105338

If you have any concerns about our products,
you can contact us on
ProductSafety@springernature.com

In case Publisher is established outside the EU,
the EU authorized representative is:
Springer Nature Customer Service Center GmbH
Europaplatz 3, 69115 Heidelberg, Germany

Printed by Libri Plureos GmbH
in Hamburg, Germany